Lecture Notes in Mathematics

Edited by A. Dold, B. Eckmann and F. Takens

1460

G. Toscani V. Boffi S. Rionero (Eds.)

Mathematical Aspects of Fluid and Plasma Dynamics

Proceedings of an International Workshop held in Salice Terme, Italy, 26–30 September 1988

Springer-Verlag
Berlin Heidelberg New York London Paris
Tokyo Hong Kong Barcelona Budapest

Editors

Giuseppe Toscani
Dipartimento di Matematica
Università de Ferrara
44100 Ferrara, Italy

Vinicio Boffi
Dipartimento di Modelli e Metodi
Matematici per le Scienze Applicate
Università di Roma "La Sapienza"
00161 Roma, Italy

Salvatore Rionero
Dipartimento di Matematica
Università degli Studi
80134 Napoli, Italy

Mathematics Subject Classification (1980): 76P05, 82A40, 76D05

ISBN 3-540-53545-4 Springer-Verlag Berlin Heidelberg New York
ISBN 0-387-53545-4 Springer-Verlag New York Berlin Heidelberg

Printed in Germany

Printing and binding: Druckhaus Beltz, Hemsbach/Bergstr.
2146/3140-543210 – Printed on acid-free paper

PREFACE

These proceedings record the main lectures given at the third International Workshop on Mathematical Aspects of Fluid and Plasma Dynamics, held in Salice Terme (Italy) from september 26 to september 30, 1988. The Workshop was sponsored by the National Council of the Research (C.N.R.), by the National Group of Mathematical Physics (G.N.F.M.), by the University of Pavia and by the Institute of Numerical Analysis of C.N.R. (I.A.N.). There were 80 participants from 8 countries. The scientific program comprised 18 main lectures and 17 contributed communications.

The main objective of the Workshop, following the first one held in Trieste, Italy, from May 30 to June 2, 1984, and the second one held in Paris-Orsay from June 24 to June 28, 1986, was to develop discussions between plasma theorists, applied mathematicians and numerical analists involved in plasma physics, fluid dynamics and kinetic theory, with emphasis on mathematical aspects of the theory. This involved a questioning of the basis equations, of considered approximations and solution methods, as well as the analysis of mathematical structures underlying physical phenomena, and the ability of these structures to represent observed experimental facts.

G.Toscani, V.Boffi, S.Rionero
Organizers and Editors

CONTENTS

MODELING INTENSE RELATIVISTIC ELECTRON BEAMS

A.M.Anile
Dipartimento di Matematica
Universita' di Catania
(Italy)

1. Introduction.

The subject of intense relativistic electron beams (IREB) is of considerable scientific and technological interest. Applications of intense relativistic electron beams include, among others, ion acceleration in linear beam geometries, plasma heating via collective instabilities, high power microwave generation and in particular the free electron laser [1]. These applications require a deep understanding of the equilibrium and stability properties of magnetically confined nonneutral plasmas that are characterized by intense self-electric and magnetic fields. It is assumed that these nonneutral plasmas are collisionless, i.e. their properties are investigated for time scales short compared with a binary collision time. Therefore atomic processes and discrete particle interactions can be neglected and collective processes dominate on the time and length scales of interest. The appropriate theoretical description then, at the microscopic level, is the kinetic description based on the Vlasov-Maxwell equations [2]. However in this framework the analysis of equilibrium configurations and of their stability (except in the simplest cases) presents serious difficulties, particularly in the non-linear regime [2]. Therefore one is forced to resort to numerical simulation techniques in order to obtain at least some qualitative results. On the other hand numerical simulation is ill-suited for a proper understanding of instabilities (which are an essential feature of IREB). As a consequence, another approach has also been followed, leading to a fluid macroscopic description of the plasma. This approach is based on the moment equations obtained from the Vlasov one, closed by some judicious approximations. This description of a nonneutral plasma is useful because of its simplicity (in particular geometric effects can be taken care of easily and stability analysis can draw on the powerful methods of fluid-dynamics stability theory). However the closure of the moment equations is well established only in the cold plasma regime, in which the stress tensor is assumed to be vanishing. How to incorporate a thermal spread in the distribution function (finite-temperature effects) is still a major problem. Several suggestions have been made in the literature [3],[4],[5],[6],[7] but we think that the subject needs further clarification. The problem of the closure of the moment equations is an old one in kinetic theory [8] and its solution depends on the physical problem under consideration. In a relativistic framework a novel feature arises, viz., that the approximation employed must be consistent with Lorentz-invariance and lead to Lorentz-invariant equations.

In this article we shall review the main attempts at building a (finite-temperature) relativistic fluid model for IREB's and examine critically their features. In particular we will stress questions regarding Lorentz-invariance.

2. Simple fluid models .

In a given inertial frame (x^i, t) one introduces the one-particle distribution function

$$f(x^i, t, p_i)$$

where $\mathbf{p}$ is the momentum of a particle related to its velocity $\mathbf{v}$ by

$$m\mathbf{v} = \frac{\mathbf{p}}{\left(1 + \frac{\mathbf{p}\cdot\mathbf{p}}{m^2}\right)^{1/2}} \qquad (1)$$

with m the particle rest-mass. The units are such that the speed of light is equal to unity.

The distribution function evolves according to the relativistic Vlasov equation [2]

$$\frac{\partial f}{\partial t} + v^j \frac{\partial f}{\partial x^j} + q\,[E_j + (v\times B)_j]\,\frac{\partial f}{\partial p_j} = 0 \qquad (2)$$

where **E** and **B** are the electric and magnetic fields, q the electric charge. Let us define the particle density n(x,t) and mean velocity **V** (x,t) by

$$n(x,t) = \int f(x,t,p)d^3p \qquad (3)$$

$$V(x,t) = \frac{1}{n}\int v f(x,t,p)d^3p \qquad (4)$$

By assuming that f(x,t,p) vanishes sufficiently fast for large momenta, from eq.(2) we obtain the particle number conservation equation

$$\frac{\partial n}{\partial t} + \nabla.(nV) = 0 \qquad (5)$$

Likewise, introducing the mean momentum

$$P(x,t) = \frac{1}{n}\int p f(x,t,p)d^3p \qquad (6)$$

multiplying eq.(2) by **p** and integrating gives

$$\frac{\partial (nP)}{\partial t} + \nabla.(nV\otimes V + \Pi) = qn\,[E + V\times B] \qquad (6)$$

where

$$\Pi = \int (v-V)\otimes(v-V) f d^3p \qquad (7)$$

is the pressure tensor.

Eq.(6) can also be rewritten as

$$\frac{\partial P}{\partial t} + V.\nabla P + \frac{1}{n}\nabla.\Pi = q\,[E + V\times B] \qquad (8)$$

Eqs. (5-7) are not a closed system. Approximations must be sought in order to close it. The simplest one corresponds to the *cold fluid model*. In this approximation one assumes that

$$f(x,p,t) = n(x,t)\delta[p-P(x,t)] \qquad (9)$$

whence

$$\int f v d^3p = nV = \frac{n\,P}{[1+ p^2/m^2]^{1/2}}$$

which gives

$$P = m\Gamma V$$

with Γ the Lorentz factor, $\Gamma = (1-V^2)^{1/2}$.

Also one has $\Pi = 0$ and the equation of motion (8) reduces to

$$\frac{\partial}{\partial t}(m\Gamma V) + V.\nabla(m\Gamma V) = q[E + V\times B] \qquad (10)$$

Remark 1

The cold fluid approximation is Lorentz-invariant although we have assumed $\Pi = 0$ in a given inertial frame. This can be seen as follows. First of all we recall that the phase-space volume element d^3p is not Lorentz-invariant, whereas $d\omega = d^3p/p^0$, with $p^0 = (p^2+m^2)^{1/2}$ is Lorentz-invariant [10]. Let p^μ be the four-vector $p^\mu = (p^0, p^i)$ and define the particle number four-flux

$$N^\mu = \int f p^\mu d\omega \qquad (11)$$

which is a four-vector. We have

$$N^0 = n \quad , \quad N^i = nV^i.$$

Also, the energy-momentum tensor $T^{\alpha\beta}$ is defined by

$$T^{\alpha\beta} = \int f p^\alpha p^\beta d\omega \qquad (12)$$

and one has

$$T^{0k} = nP^k \quad , \qquad T^{jk} = nV^jP^k + \Pi^{jk}$$

and the LHS of eq.(6) is the four-divergence $\partial_\alpha T^{\alpha k}$.

Now, for a cold fluid, we have also $T^{00} = nP^0$ where $P^0 = (P^2+m^2)^{1/2}$.

Then by introducing the rest-mass density $\mu = nm/\Gamma$, it is easily seen that, for a cold fluid,

$$T^{\alpha\beta} = \mu\, U^{\alpha} U^{\beta}$$

where $U^{\alpha} = P^{\alpha}/m$. Also, $N^{\alpha} = \mu\, U^{\alpha}$. Therefore a cold fluid can be described covariantly by the particle number conservation

$$\nabla_{\alpha} N^{\alpha} = 0 \tag{13}$$

and the energy-momentum conservation laws

$$\nabla_{\alpha} T^{\alpha\beta} = \frac{q}{m} N_{\alpha} F^{\beta\alpha} \tag{14}$$

where ∇_{α} denotes the covariant derivative and $F^{\alpha\beta}$ the electromagnetic field tensor, $E_i = F_{io}$, $B_1 = F_{23}$, $B_2 = F_{23}$, $B_3 = F_{12}$. These equations must then supplemented by Maxwell's ones.

Cold plasma equilibrium configurations and their stability have been extensively studied in the cold fluid approximation. One of the simplest examples of a beam steady flow is the axisymmetric Bennett pinch, where the repulsive electrostatic forces are balanced by the magnetic forces in the radial direction. The geometry is as in Fig.1. The following assumptions are made :

i) the plasma is infinite and uniform in the z-direction and furthermore $n=n(r)$, $\mathbf{V}=\mathbf{V}(r)$;

ii) the electrons have no mean motion in the radial and azimuthal direction, $V_r = V_\theta = 0$ and the axial velocity is assumed constant, $V_z =$ constant ;

iii) the ions are taken to form a static, partially neutralizing background with density $n_i(r) = fn(r)$ where f=const. is the fractional neutralization.

Then the continuity equation (5) is automatically satisfied. Maxwell's equations are

$$\nabla \times \mathbf{E} = -\frac{\partial \mathbf{B}}{\partial t}, \qquad \nabla \times \mathbf{B} = -4\pi e n \mathbf{V} + \frac{\partial \mathbf{E}}{\partial t}$$

$$\nabla \cdot \mathbf{E} = -4\pi n e(f-1), \qquad \nabla \cdot \mathbf{B} = 0$$

where e is the absolute value of the electron charge. The electric and magnetic fields are then

$$E = E_r e_r \quad , \quad B = B_\theta e_\theta$$

with e_r, e_θ unit vectors in the radial and azimuthal directions and

$$\frac{1}{r}\frac{d}{dr}(rE_r) = -4\pi en(1-f) \tag{15}$$

$$\frac{1}{r}\frac{d}{dr}(rB_\theta) = -4\pi enV_z \tag{16}$$

The radial component of eq.(8), for a cold fluid ($\Pi = 0$) gives then

$$E_r = V_z B_\Theta = \beta B_\theta$$

where $\beta = V_z$ (in units of the light speed), and from (15-16) it follows

$$\beta^2 = 1-f \tag{17}$$

Then, assigning $n=n(r)$, β, f the complete solution is obtained by integrating eq. (15) . Therefore, for a cold beam (satisfying eq.17) any density profile is possible in principle.

The generalization of this solution to the case of finite temperature leads to several problems. The customary way of including thermal effects is to assume [2] :

i) that the pressure tensor is isotropic, i.e.

$$\Pi_{jk} = \Pi\,\delta_{jk}$$

ii) the pressure Π obeys the ideal gas law

$$\Pi = nk_B T$$

T being the absolute temperature and k_B the Boltzmann constant ;

iii) the beam is isothermal, T = const.

Eqs.(15-16) can be integrated immediately, giving

$$E_r = - \frac{4\pi en}{r} (1-f) \int_0^r n(r')r'dr' \tag{18}$$

$$B_\theta = - \frac{4\pi en}{r} \beta \int_0^r n(r')r'dr' \tag{19}$$

The radial component of eq.(8) then yields

$$k_B T \frac{1}{n} \frac{dn}{dr} = \frac{4\pi e^2}{r} (1-f-\beta^2) \int_0^r r'n(r')dr' \tag{20}$$

The solution of this equation is the Bennett profile

$$n(r) = \frac{n(0)}{[1 + r^2/a^2]^2} \tag{21}$$

where

$$a^2 = \frac{8 \lambda_D^2}{\beta^2 - (1-f)} \tag{21'}$$

with

$$\lambda_D = [\frac{k_B T}{4\pi e^2 n(0)}]^{1/2}$$

the Debye length .Notice that one must have $1-f < \beta^2$.

There are several unsatisfactory features with this analysis :

i) the assumption of an isotropic pressure tensor in a given inertial frame is not Lorentz-invariant. A more sensible assumption would be to assume *isotropy in the local rest frame of the fluid*;

ii) likewise, the ideal gas law can be justified only in the local rest-frame ;

iii) the meaning of isothermality in a frame which is not the local rest one is dubious .

Another very important difficulty is the following :

iv) the cold fluid limit seems to be singular. In fact, by making $1-f$ approach β^2 , we should recover the cold fluid limit. However, in this limit, the temperature T does not appear. Therefore, for physical consistency, an important link must exist between T, f and β , which the present theory is not capable of determining.

The first three difficulties can be removed by working with

covariant fluid models . The last difficulty is of a more fundamental nature and requires the consideration of higher order fluid models (including higher order moments). Following Toepfer[3] we describe the beam as a relativistic perfect fluid obeying the conservation equations (13-14) with the energy-momentum

$$T^{\alpha\beta} = (\mu+p)U^{\alpha}U^{\beta}+pg^{\alpha\beta} \qquad (22)$$

where U^{α} is the fluid's four-velocity, μ and p the rest-frame total energy-density and pressure, $g^{\alpha\beta}$ the metric tensor. For relativistic perfect fluids there is a well established covariant theory [9][10][11]. We assume the ideal gas state equation in the local rest-frame, which has a kinetic justification, i.e. the Synge state equation[12],

$$\mu = n_o m \left(G - \frac{1}{\xi}\right)$$

$$p = mn_o/\xi \qquad (23)$$

with n_o the local rest-frame particle density, $\xi = m/k_B T_o$, T_o the rest-frame temperature and

$$G = K_3(\xi)/K_2(\xi)$$

$K_n(\xi)$ being the modified Bessel function of order n .

From eq. (14) by contracting with the projection tensor

$$h^{\mu\nu} = g^{\mu\nu}+U^{\mu}U^{\nu}$$

we obtain the momentum balance equation [10]

$$(\mu+p)U^{\mu}\nabla_{\mu}U_{\alpha} + h_{\alpha}{}^{\nu}\nabla_{\nu}p = -en_o F_{\alpha\mu}U^{\mu} \qquad (24)$$

For the Bennett flow it is convenient to exploit Lorentz-covariance and work in the inertial rest-frame of the fluid, in which $U^{\alpha}=(1,0,0,0)$.

The radial component of eq.(24) gives

$$\frac{dp}{dr} = -en_o E^o_r \qquad (25)$$

where

$$E_r^o = -\frac{4\pi e}{r}n_o(1-f_o)\int_o^r n_o(r')r'dr'$$

with f_o the neutralization factor in the beam rest-frame. Since

$$p = nk_B T_o$$

we obtain

$$k_B T_o \frac{1}{n_o}\frac{dn_o}{dr} = -\frac{4\pi e^2}{r}(f_o-1)\int_o^r n_o(r')r'dr'$$

whence

$$n_o(r) = \frac{n_o(0)}{[1 + r^2/a_o^2]^{1/2}} \tag{26}$$

where

$$a_o^2 = \frac{8\lambda_{OD}^2}{f_o-1}$$

with λ_{OD} the rest-frame Debye length.
Now, $f_o = n_i/n_o$, n_i being the ion density in the beam rest-frame, whereas $f = n_{io}/n$, n_{io} being the local rest-frame ion density. Hence $f_o = \Gamma^2 f$ and

$$a_o^2 = \frac{8k_B T_o/\Gamma}{[\beta^2-(1-f)]4\pi e^2 n(0)} \tag{27}$$

with $n(0)=\Gamma n_o(0)$. Then the expression (27) coincides with (21') provided $T = T_o/\Gamma$, which is a reasonable relativistic transformation law for temperature [12].

In order to answer question iv) (the singular cold fluid limit starting from the finite temperature Bennett solution) it is necessary to view the ideal fluid model as a first step in an approximation scheme leading to higher order fluid models.

3. Covariant warm fluid models.

Let u^μ denote the four-velocity of a particle, which, in inertial coordinates (x^μ), writes

$$u^\mu = \Gamma(v^i, 1)$$

where $\Gamma = (1 - v^i v_i)^{-1/2}$ is the Lorentz factor and v^i the three velocity. Let $u^i = \Gamma v^i$ and introduce the one particle invariant distribution function $f(x^\mu, u^i)$. Then the Lorentz covariant Vlasov equation writes [6]

$$u^\mu \frac{\partial f}{\partial x^\mu} + \frac{e}{m} F^{i\mu} u_\mu \frac{\partial f}{\partial u^i} = 0 \tag{28}$$

with e the electric charge, m the particle mass .

This equation can be written in a manifestly covariant form by introducing the function

$$F(x^\mu, u^\mu) = f(x^\mu, u^i)$$

defined on the hyperboloid $u_\mu u^\mu = -1$, for which eq. (28) writes

$$u^\mu \frac{\partial F}{\partial x^\mu} + \frac{e}{m} F^{\alpha\mu} u_\mu \frac{\partial F}{\partial u^\alpha} = 0$$

The following quantities related to the moments of the distribution function f can be defined

$$h = \int f\omega \tag{29}$$

$$h w^\mu = \int f u^\mu \omega \tag{30}$$

$$\Theta^{\mu\nu} = \int f(u^\mu - w^\mu)(u^\nu - w^\nu)\omega \tag{31}$$

$$S^{\mu\nu\lambda} = \int f(u^\mu - w^\mu)(u^\nu - w^\nu)(u^\lambda - w^\lambda)\omega \tag{32}$$

where $\omega = \frac{du^1 du^2 du^3}{u^0}$ is the invariant measure on the velocity space. Notice that the scalar h is different from the previously introduced quantity n_0 .The quantity h can be interpreted as an invariant particle density, w^μ as an average four-velocity.

From the definitions (31) and (32) the following relationships are obtained

$$\Theta^{\mu}{}_{\mu} = - h\,(1 + w^{\mu} w_{\mu}) \tag{33}$$

$$S^{\mu\nu}{}_{\nu} = - 2\,\Theta^{\mu\nu} w_{\nu} \tag{34}$$

By taking the moments of the Vlasov equation (28), assuming that f vanishes sufficiently fast at infinity in the velocity space, we obtain

$$\frac{\partial h w^{\mu}}{\partial x^{\mu}} = 0 \tag{35}$$

$$\frac{\partial}{\partial x^{\mu}} (h\, w^{\mu} w^{\nu} + \Theta^{\mu\nu}) = \frac{e}{m} h\, F^{\mu\nu} w_{\mu} \tag{36}$$

$$\frac{\partial}{\partial x^{\mu}} (h\, w^{\mu} w^{\nu} w^{\alpha} + w^{\mu}\Theta^{\nu\alpha} + w^{\nu}\Theta^{\mu\alpha} + w^{\alpha}\Theta^{\mu\nu} + S^{\mu\nu\alpha}) =$$

$$= \frac{e}{m} (F^{\nu\mu}\Theta^{\alpha}{}_{\mu} + F^{\alpha\mu}\Theta^{\nu}{}_{\mu} + F^{\nu\mu} w_{\mu} h w^{\alpha} + F^{\alpha\mu} w_{\mu} h w^{\nu}) \tag{37}$$

The set of equations (35),(36),(37) obviously is not a closed system. In order to close it one must introduce constitutive equations for $S^{\mu\nu\lambda}$ on the basis of some approximation scheme and determine their range of validity .

The basic physical approximation will be that the spread in momentum space is very small, expressing the concept of warm plasma. More precisely we introduce the following quantity

$$\varepsilon^2 = -(1 + w^{\mu} w_{\mu}) \tag{38}$$

and define a warm plasma by the requirements : i) $\varepsilon \ll 1$ ii) in the local rest frame of the plasma, in which

$$w^{\mu} = (\bar{w}^{0},0,0,0\ ) \quad \text{with} \quad \bar{w}^{0} = \sqrt{1+\varepsilon^2},$$

the following representation holds

$$\frac{\bar{\Theta}^{ij}}{h} = [\bar{K}^{ij} + \frac{1}{3} (1 - \bar{K}^{l}{}_{l})\delta^{ij}]\varepsilon^2 + \frac{1}{3} \bar{K}^{00}\, \delta^{ij} \varepsilon^4$$

$$\frac{\bar{\theta}^{io}}{h} = \bar{K}^{io}\varepsilon^3 , \qquad \frac{\bar{\Theta}^{oo}}{h} = \bar{K}^{oo}\varepsilon^4$$

$$\frac{\bar{S}^{ijk}}{h} = [\bar{K}^{ijk} - \frac{3}{5}\delta^{(ij}\,\bar{K}^{k)l}{}_l]\varepsilon^3 + \frac{6}{5}\delta^{(ij}\bar{K}^{k)o}\varepsilon^3\sqrt{1+\varepsilon^2} +$$
$$+\frac{3}{5}\delta^{(ij}\bar{K}^{k)oo}\varepsilon^5 \tag{39}$$

$$\frac{\bar{S}^{ijo}}{h} = [\bar{K}^{ij4} - \frac{1}{3}\bar{K}^{ol}{}_l\delta^{ij}]\varepsilon^4 + \frac{2}{3}\delta^{ij}\bar{K}^{oo}\varepsilon^4\sqrt{1+\varepsilon^2} +$$
$$+ \frac{1}{3}\delta^{ij}\bar{K}^{ooo}\varepsilon^6$$

$$\frac{\bar{S}^{ioo}}{h} = \bar{K}^{ioo}\varepsilon^5, \qquad \frac{\bar{S}^{ooo}}{h} = \bar{K}^{ooo}\varepsilon^6$$

where overbars denote components in the plasma local rest frame and $K^{\mu\nu}$, $K^{\mu\nu\lambda}$ are dimensionless tensors at most of order zero in ε, such that $K^{\mu\nu} = K^{(\mu\nu)}$, $K^{\mu\nu\lambda} = K^{(\mu\nu\lambda)}$. It is easy to check that the above ordering satisfies the constraints (6) and (7) exactly.

Notice that for a distribution function of the type $f = h\,\delta(u^\mu - w^\mu)$, leading to a cold fluid, one has $\varepsilon = 0$. The assumptions (39) can be justified for a nearly thermal distribution function as $T_o \to 0$ [13].

The ordering (39) suggests that the moment equations (35-36-37) could be closed by systematically neglecting terms of higher order in ε. The cold fluid models (eqs.13-14) can be obtained by neglecting terms of order ε^2 and keeping those of order O(1) . In fact in this case one puts $\theta^{\alpha\beta} = 0$. A higher order warm fluid model would be obtained by neglecting $S^{\mu\nu\alpha}$ altogether, and this leads to the model of Amendt and Weitzner [6].

However the closure method of Amendt and Weitzner [6] does not satisfy the constraint (34) exactly.

. A better approach would be to look for $S^{\mu\nu\alpha}$ as function of the previous moments h, w^α, $\theta^{\mu\nu}$ such that it *satisfies the constraint* (34) exactly and *preserves the physical ordering* (39).

Therefore, we shall look for a "constitutive function"

$$S^{\mu\nu\alpha} = \hat{S}^{\mu\nu\alpha}(h,\, w^\beta,\, \theta^{\alpha\beta})$$

such that:

i) the constraint (34) is satisfied exactly,

ii) the ordering (39) is preserved.

Anile and Pennisi [7][13] have shown that entropy arguments lead to the following constitutive function which satisfies the requirements i) ii) :

$$S^{\alpha\beta\gamma}=-2(1+w^{\tau}w_{\tau})(-w^{\delta}w_{\delta})^{-1/2}\partial^{\mu\nu}\left[h^{\langle\alpha}{}_{\mu}h^{\beta}{}_{\nu}-\frac{1}{3}h^{\langle\alpha\beta}h_{\mu\nu}\right]w^{\gamma\rangle}+$$

$$+4(1+w^{\tau}w_{\tau})(-w^{\delta}w_{\delta})^{-3/2}\left[\frac{1}{5}h^{\langle\alpha\beta}(-w^{\lambda}w_{\lambda})+w^{\langle\alpha}w^{\beta}\right]h^{\gamma\rangle}{}_{\mu}\partial^{\mu\nu}w_{\nu}-$$

$$-(1+w^{\tau}w_{\tau})(-w^{\delta}w_{\delta})^{-3/2}\left\{\frac{h}{9}(1+w^{\lambda}w_{\lambda})^{2}+\left[\theta^{\mu\nu}w_{\mu}w_{\nu}(-w^{\lambda}w_{\lambda})^{-1}-\right.\right.$$

$$\left.\left.-\frac{h}{6}(1+w^{\lambda}w_{\lambda})^{2}\right]\left(\frac{2}{15}\alpha_{1}+2\right)\right\}(w^{\langle\alpha}w^{\beta}-w^{\sigma}w_{\sigma}h^{\langle\alpha\beta})w^{\gamma\rangle}-$$

$$-2h^{\langle\alpha\beta}\left(\frac{w^{\gamma\rangle}w_{\delta}}{w^{\lambda}w_{\lambda}}-\frac{3}{5}h^{\gamma\rangle\delta}\right)\theta_{\delta\mu}\,w^{\mu} \tag{40}$$

where α_1 is an arbitrary function of h and $h_{\alpha\beta}$ is the projection tensor.

With this choice the moment equations (35-36-37) form a hyperbolic system verifying a supplementary conservation law (in the form of an entropy principle [14]) and are equivalent to a symmetric hyperbolic system. To the order ε^3 the complicated expression (40) can be approximated by the simple form

$$S^{\alpha\beta\gamma} = \frac{6}{5}h^{\langle\alpha\beta}h^{\gamma\rangle\delta}\,\partial_{\delta\mu}w^{\mu} + O(\varepsilon^4) \tag{41}$$

Amendt and Weitzner [15] have studied the Bennett pinch solution in the framework of thei warm fluid model. Since the cold fluid model can be viewed as a first step in the higher order warm fluid models, by imposing the correct ordering on $S^{\alpha\beta\gamma}$, they are able to show that the rest-frame beam radius must be of order of the beam skin depth and this observation solves the problem mentioned in Sec.2, iv) , regarding the singular limit of the cold fluid model.

REFERENCES

(1) R.B.Miller. *Intense Charged Particle Beams*, Plenum Press, N.Y. (1985).

(2) R.C.Davidson, *Theory of Nonneutral Plasmas*, Benjamin, Reading, Mass. (1974).

(2) A.J.Toepfer, Phys.Rev.A3,1444 ,(1971).

(4) J.G. Siambis, Phys. Fluids, 22, 1372, (1979).

(5) W.A Newcomb , Phys. Fluids, 25, 846 ,(1982).

(6) P.Amendt and H.Weitzner, Phys. Fluids, 28, 949, (1985).

(7) Anile,A.M. and Pennisi,S. An improved relativistic warm plasma model, to be published in A.M.Anile and Y.Choquet-Bruhat (Eds.), *Relativistic Fluid Dynamics*, Proceedings of the C.I.M.E Symposium on Relativistic Fluid Dynamics, (Noto, 1987) .

(8) H.Grad, Comm.Pure and Appl.Math., vol.V, 257 (1952)

(9) A.Lichnerowicz, *Relativistic Hydrodynamics and Magnetohydrodynamics*, Benjamin, N.Y. (1967) P.Amendt and H.Weitzner Phys.Fluids 30,

(10) J.L.Synge, *Relativity: The Special Theory*, North Holland, Amsterdam (1956).

(11) A.M.Anile, *Relativistic Fluids and Magnetofluids* , Cambridge University Press, Cambridge (1989).

(12) J.L.Synge, *The Relativistic Gas*, North Holland, Amsterdam, (1957)

1814 (1987)

(13) A.M.Anile & S.Pennisi, *Fluid models for relativistic electron beams*, University of Catania, Preprint (1988).

(14) I-Shih Liu,I.Müller and T.Ruggeri, Annals of Physics, 159, 191, (1986).

(15) P.Amendt & H.Weitzner, Phys.Fluids 30, 1814 (1987).

(16) W.G.Dixon, *Special Relativity*, Cambridge University Press, Cambridge (1978).

ON THE ASYMPTOTIC THEORY OF THE BOLTZMANN AND ENSKOG EQUATIONS
A RIGOROUS H-THEOREM FOR THE ENSKOG EQUATION

N. Bellomo and M. Lachowicz*
Department of Mathematics, Politecnico of Torino, Italy
*On leave from Depart. Mathematics, University of Warsaw, Poland

Abstract: This paper developes the mathematical theory of the asymptotic equivalence, initiated by the authors in a previous paper [3], between the Boltzmann and the Enskog equations referred to the solutions to the initial value problem when the radius of the hard spheres in the Enskog equation tends to zero. This paper deals with an H-theorem for the Enskog equation and proves an asymptotic equivalence result which states that the Liapunov functional proposed by Polewczak [13], referred to the solution of the initial value problem for the Enskog equation, is monotone decreasing in time and tends, when the radius of the spheres goes to zero, to the H function referred to the Boltzmann equation.

1. On the Theory of the Asymptotic Equivalence

One of the fundamental models of the nonlinear kinetic theory of gases is the Enskog equation proposed by Enskog [7] under somehow heuristic arguments. The model has been afterwords revised by various authors and in particular by Resibois [15] according to a detailled analysis mainly oriented to make the model consistent with the irreversible thermodynamics.

The model proposed by Ensog introduces two important modifications with respect to the Boltzmann equation, both related to the fact that gas particles are not, in physical reality, point masses, but have a finite dimension. In other words the gas particles are modelled as spheres with radius σ.

This feature is taken into account assuming that the collision between gas particles involves pairs of particles with their centers placed at a distance equal to 2σ. Moreover the collision frequency is increased by a factor Y which in the original Enskog equation is a function of the gas density at the contact point of the two colliding spheres and in the revised Enskog equation is a functional of the local gas density. The equation still neglects multiple collisions.

It is not surprising that this equation has originated several discussions. In fact this model introduces some interesting modifications of the Boltzmann equation. These modifications are certainly in the direction of making the Enskog model more consistent with physical reality then the Boltzmann one.

On the other hand a rigorous derivation of the Enskog equation appears even more difficult to justify, on a rigorous basis, than the one of the Boltzmann equation. In addition the Enskog equation is not even consistent, in its original formulation, with irreversible thermodynamics.

Keeping in mind these remarks and having in the background the general analysis developed by Lebowitz et al. [10], the authors of this paper have originated an asymptotic theory for the Enskog equation essentially oriented to show the equivalence between the Boltzmann and the Enskog model [3]. Where the term "equivalence" is here used to indicate that the solutions to suitable initial value problems referred to both equations tends asymptotically to the same value (if, of course, the initial datum is the same) when the radius of the hard spheres in the Enskog equation tends to zero.

This analysis has shown, in particular, that the initial value problem for the Enskog equation always has a solution when the same problem for the Boltzmann equation has a smooth solution and if the radius of the spheres is below a critical value. Moreover, papers [8, 9] have studied various aspects of the hydrodynamic limit for the Enskog equation when the Knudsen number for the gas particles tends to zero jointly with the radius of the particles.

It is worth mentioning that paper [3] contains a proof of global existence of classical solutions to the initial value problem for the Enskog equation and that a similar results has been obtained by Polewczak [12]. Whereas papers [8, 9] extend to the Enskog equation the theorems available for the Boltzmann equation on the hydrodynamic limit. These asymptotic theories are now addressed to several other directions and, in particular, to develope the powerfull theorem by Di Perna and Lions [5] on the existence, without uniqueness, of the solution of the initial value problem for the Enskog equation for large initial datum [1].

This paper developes this asymptotic theory towards the analysis of a rigorous H-theorem for the Enskog equation with the aim to prove an asymptotic equivalence theorem between the H-function for the Boltzmann equation and the Resibois-Polewczak Liapunov functional for the Enskog equation. The proof of such a theorem refers to a well prescribed initial value problem and is contained in the third section of this paper which follows a general discussion on the H-theorem and irreversibility provided in the second section of this paper.

2. On the H-Theorem and Irreversibility

Let us introduce, to be specific and avoid too general arguments, the two mathematical models we are dealing with. Consider then the Boltzmann and the Enskog equation written, with reference to [3] as well as to Chap.IV of [4], in the following dimensionless form

$$\partial f_B/\partial t + v\cdot\underline{\nabla}_x f_B = J(f_B,f_B) \tag{1}$$

$$\partial f_E/\partial t + v\cdot\underline{\nabla}_x f_E = E_\sigma(f_E;f_E,f_E) \tag{2}$$

with initial data

$$f_E(0,x,v) = f_B(0,x,v) = F(x,v) \tag{3}$$

where the Knudsen number has been put, for simplicity, equal to one. f_B and f_E are dimensionless distribution function, t, x and v the dependent variables; $t\in R$, $x\in R^3$ and $v\in R^3$ are (dimensionless) time, position and velocity, respectively. The parameter σ is the dimensionless diameter of the particles.

The collision operators J and E_σ characterizing eqs.(1,2) can be written as follows

$$J(f,f)(x,v)=\int_{R^3\times S^2}\{f(x,v_1')f(x,v') - f(x,v_1)f(x,v)\}\phi((v_1-v)\cdot\nu)d\nu dv_1 \tag{4}$$

$$E_\sigma(f;f,f)(x,v)=\int_{R^3\times S^2}\{Y^+(f;\sigma)(x,\nu)f(x+\sigma\nu,v_1')f(x,v') -$$

$$- Y^-(f;\sigma)(x,\nu)f(x-\sigma\nu,v_1)f(x,v)\}\phi((v_1-v)\cdot\nu)d\nu dv_1 \tag{5}$$

where

$$S^2 = \{\nu \in R^3 : |\nu| = 1\} , \qquad \phi(y) = \max\{0,y\} \tag{6}$$

$Y^{\pm}$ are the pair correlation functions and the definitions of the post collisional velocities v' and v_1' are the classical ones [4].

If now we consider the functions along the free-streaming trajectories

$$f^{\#}(t,x,v) = f(t,x+vt,v) \tag{7}$$

the "mild" form of equations (1,2) can be written

$$f_B^{\#} = F + (\mathscr{U}\, J(f_B,f_B))^{\#} \tag{8}$$

and

$$f_E^{\#} = F + (\mathscr{U}\, E_\sigma(f_E;f_E,f_E))^{\#} \tag{9}$$

where the operator $\mathscr{U}$ is defined by

$$\mathscr{U}f(t,x,v) = \int_0^t f(s,x-v(t-s),v)ds \tag{10}$$

We need now, in order to deal with the problem which has been announced in the introduction, some function spaces. Therefore if W_α is the function defined by $W_\alpha(y)=(1+y^2)^{\frac{1}{2}\alpha}$ and $C_b^n(X)$ is the space of all real functions continuous and bounded with all their derivatives of order $|\gamma| < n$, on the space X, then we can define the space

$$B^n_{p,k} = \{f \in C_b^0(R^3\times R^3): (\partial^{|\gamma|}f(x,v)/\partial x^\gamma)W_p(|x|)W_k(|v|) \in C_b^0(R^3\times R^3)$$

$$\text{for each multi-index } \gamma \text{ such that } |\gamma| \leq n\} \tag{11}$$

which can be endowed with the norm

$$|||f|||^n_{p,k} = \sup_{\substack{|\gamma|\leq n \\ (x,v)\in R^3\times R^3}} |(\partial^{|\gamma|}f(x,v)/\partial x^\gamma)W_p(|x|)W_k(|v|)| \tag{12}$$

in addition we can also define the space

$$\mathcal{B}^n_{p,k} = \{f \in C^0_b(R_+ \times R^3 \times R^3):$$

$$(\partial^{|\gamma|} f/\partial x^\gamma)^{\#}(t,x,v) W_p(|x|) W_k(|v|) \in C^0_b(R_+ \times R^3 \times R^3),$$

$$\text{for all multi-index } \gamma \text{ such that } |\gamma| \le n\} \tag{13}$$

which can be endowed with the norm

$$|||f|||^n_{p,k} = \sup_{t \ge 0} \| f^{\#}(t) \|^n_{p,k} \tag{14}$$

Consider now the classical H-functional referred to the Boltzmann equation

$$H[f](t) = \int_{R^3 \times R^3} f(t,x,v) \ln f(t,x,v) dx dv \tag{15}$$

Moreover, consider the following two conditions

$$\forall t \ge 0 : \quad (f \ln f)(t) \in L_1(R^3 \times R^3) \tag{16}$$

and

$$t \uparrow \;\Rightarrow\; H[f] \downarrow \tag{17}$$

It is well known in the literature [14] that a formal H-theorem holds for the Boltzmann equation when conditions (17) is "formally" satisfied.

Such a Theorem can be regarded as a rigorous one if both conditions (16,17) can be proven at least for suitable initial conditions. Such a project has been realized, for the Boltzmann equation, in paper [16] which will become the starting point for the analysis developed in this paper.

Referring to the function spaces defined in (11, 13) Toscani's theorem can be rewritten as follows

<u>THEOREM 1</u> (Toscani [16])

i) Let $p > 1$ and $k > 3$, then there exists a constant c_1 such that <u>if</u>

$$0 \le F \quad \text{and} \quad \| F \|^{0}_{p,k} \le c_1 \tag{18}$$

<u>then</u> problem (8) has a unique nonnegative solution f_B in $\mathcal{B}^{0}_{p,k}$.

ii) Let $p > 3$ and $k > 7$, then if condition (18) is satisfied the following conditions hold

$$(f_B \ln f_B)(t) \in L_1(R^3 \times R^3) \text{ for all } t \ge 0 \tag{19}$$

and

$$H[f_B](t_2) \le H[f_B](t_1) \quad \text{for} \quad t_2 \ge t_1 \ge 0 \tag{20}$$

see also [3], Chap.II.

The possibility of deriving this kind of theorem is a fundamental step in the validation of specific models of the kinetic theory of gases. Unfortunately the original Enskog equation (classified in [2] as the"standard Enskog equation") did not posses property (17) even at a formal level. Therefore some revisions of the standard Enskog equation have been necessary as documented in papers [11, 15, 17, 18] which have provided several modifications to the original equation.

In particular, Resibois [15] has provided a "modified Enskog equation" (classified in [2], as well as by other authors, as the "revised Enskog equation") essentially obtained modifying both the term Y and the H-functional itself in order to fulfill condition (17).

The limit of the analysis by Resibois consists in the fact that both the term Y and the Liapunov functional were provided at a formal level without a detailled formulation. However, such a topic has been succesfully developed by Polewczak [13] who suggested the following expression for the Liapunov functional

$$H^*[f](t) = H[f](t) - P^*[f](t) \tag{21}$$

where the term H was already defined in (15) and the term P^* is defined by

$$P*[f](t) = (1/2)\int_0^t P[f](s)ds \tag{22}$$

where

$$P[f](s) = \int_{(R^3)^3 \times S^2} G[f](s,x,v,v_1,\nu)d\nu dv dv_1 dx \tag{23a}$$

and

$$G[f](s,x,v,v_1,\nu) = \{Y^+(f;\sigma)(s,x,\nu)f(s,x+\sigma\nu,v_1) -$$

$$- Y^-((f;\sigma)(s,x,\nu)f(s,x-\nu\sigma,v_1)\}f(s,x,v)\phi((v_1-v)\cdot\nu) \tag{23b}$$

It has been shown by Polewczak [13] that condition (17), "formally", holds for the functional (18). A rigorous proof needs to show that conditions (16,17) both holds for suitable initial data. An asymptotic equivalence theorem needs to show, in addition, that $\sigma \longrightarrow 0 \Rightarrow H^* \longrightarrow H$. This is the topic dealt with in the next section.

It is important to mention, in order to put into the correct light the analysis developed in this paper, that the analysis provided by Polewczak has been developed having in the background the powerfull theorem proved by Di Perna and Lions [5] for the Boltzmann equation. The aim of paper [13] consist in proving an analogous existence result for the Enskog equation, namely an existence theorem (without uniqueness) for large inital data. On the other hand, the analysis developed in this paper is such that the existence proof can be joined to a uniqueness result. Then one has to pay the price of smallness assumptions which are not needed in the case of paper [5].

3. Asymptotic Equivalence and H-Theorem for the Enskog Equation

Consider now the revised Enskog equation proposed by Resibois [15] and the H functional proposed by Polewczak [13] with the aim of proving the theorem already announced in the preceding section. With this in mind we first need precising some properties of the term Y.

We assume then that $Y^{\pm}$ is a function of the local density

$$\rho(t, x) = \int_{R^3} f(t,x,v)dv$$

at the points x and $x \pm \sigma\nu$, that is

A.1. $Y^{\pm}(f;\sigma)(t,x,\nu) = Y(\rho(t,x), \rho(t,x\pm\sigma\nu))$

and is symmetric

A.2. $Y(\rho_1,\rho_2) = Y(\rho_2,\rho_1)$.

Moreover we need the following additional assumptions

A.3 $Y^{\pm}(0;\sigma) = 1$ for all $\sigma \geq 0$,

A.4 $\forall\sigma \geq 0, \forall f_1, f_2 \in D^*_{p,k}: 0 \leq f_1 \leq f_2 \Longrightarrow Y^{\pm}(f_1;\sigma) \leq Y^{\pm}(f_2;\sigma)$

A.5 $\forall\sigma \geq 0, \forall f_1, f_2 \in D^*_{p,k}$:

$$|Y^{\pm}(f_1;\sigma)(t_1,x_1,\nu_1) - Y^{\pm}(f_2;\sigma)(t_2,x_2,\nu_2)| \leq$$

$$< \zeta(\sigma)\{ \int |f_1(t_1,x_1,v) - f_2(t_2,x_2,v)|dv +$$

$$+ \int |f_1(t_1,x_1+\sigma\nu_1,v) - f_2(t_2,x_2+\sigma\nu_2,v)|dv\}$$

where $\zeta(\sigma) \to 0$ for $\sigma \to 0$.

and $D^*_{p,k} \subset \mathcal{B}^o_{p,k}$ is a subset of functions corresponding, in the terms specified in [3], to densities which are not closed to the condensation density for which Y tends to infinity.

We need now recalling the main result of paper [3] which will be used in what follows.

<u>THEOREM 2</u> (Bellomo and Lachowicz [3])

i) Let $p > 1$ and $k > 3$ and let, in addition, condition (18) be satisfied. Then if $F \in B^n_{p,k}$ with $n > 1$, one has

$f_B \in \mathcal{B}^n_{p,k}$

where f_B is the solution of eq.(8) in the terms stated by Theorem 1, item i).

ii) Let $p > 1$ and $k > 3$ and let condition (18) be satisfied. Suppose, in addition, that $F \in B^1_{p,k}$ and Assumptions A.3, A.4 and A.5 are satisfied. Then there exists a constant $\sigma_o > 0$ such that if $0 < \sigma < \sigma_o$ then problem (9) has a unique solution f_E in $\mathring{\mathcal{B}}_{p,k}$ and, in addition f_B, $f_E \in D^*$ such that

$$|||f_E - f_B|||^{\circ}_{p,k} \to 0 \quad \text{as} \quad \sigma \to 0 \tag{24}$$

where f_B is the corresponding solution of problem (8) in the space $\mathcal{B}^1_{p,k}$.

Remark 1: The proof of Theorem 2 has been obtained in [3] with $\zeta(\sigma) = O(\sigma)$. However all proof can be repeated with decay as $\max\{\sigma, \zeta(\sigma)\}$.

After these preliminaries we can now state the main result of this paper

THEOREM 3

Let $p > 3$ and $k > 7$, $F \in B^1_{p,k}$ and condition (18) be satisfied. Moreover suppose that assumptions A.1-5 are satisfied and $0 < \sigma < \sigma_o$ where σ_o has been introduced in Theorem 2. Then, if f_B and f_E are the solutions of the initial value problem for the Boltzmann and Enskog equations, respectively, in the terms stated by Theorem 2, the following conditions hold

$$(f_E \ln f_E)(t) \in L_1(R^3 \times R^3) \qquad , \forall t \geq 0, \tag{25}$$

$$G[f_E](t) \in L_1((R^3)^3 \times S^2) \quad , \forall t \geq 0, \tag{26}$$

$$H^*[f_E](t_2) \leq H^*[f_E](t_1) \quad \text{for} \quad 0 \leq t_1 \leq t_2 < \infty \tag{27}$$

and, in addition

$$\lim_{\sigma \to 0} H^*[f_E](t) = H[f_B](t) \quad , \quad \forall t \in [0, \infty) \tag{28}$$

Proof: The proof of Theorem 3 is in three steps: first conditions (25, 26) are proven, then the second step deals with inequality (27), for special initial data, and finally inequality (27) in full generality and the asymptotic equivalence condition (28) are proven in the last step.

Step 1: Theorem 2 and, in particular, the estimate (12) of ref. [16] guarantee that $f_E \ln f_E(t)$ is uniformly bounded, with respect to t, by a function in $L_1(R^3 x R^3)$. Moreover, considering that

$$H[f_E] = H[f_E^{\#}] \tag{29}$$

Then condition (25) is satisfied. Moreover, by the Lebesgue dominated convergence theorem one has

$$H[f_E] \in C_b^o([0, \infty)) \tag{30}$$

Moreover, according to Theorem 2, item ii), and to Assumptions A.3-5, one has

$$|G[f_E](t,x+vt,v_1,\nu)| < \text{const.}(|||\overset{o}{f}|||^2_{p,k})(1+\text{const.}|||\overset{o}{f}|||_{p,k})$$

$$\{W_{-p}(|x-t(v_1-v)+\sigma\nu|) + W_{-p}(|x-t(v_1-v)-\sigma\nu|)$$

$$W_{-p}(|x|)W_{-k}(|v_1|)W_{-k}(|v|)\phi((v_1-v)\cdot\nu) \tag{31}$$

Consequently, Lemma 1 of [3] and Theorem 2 yield

$$|G[f_E](t,x+tv,v,v_1,\nu)| \leq \text{const.}\ |v_1-v|W_{-2}(|x-t(v_1-v|)$$

$$W_{-p}(|x|)W_{-k}(|v|)W_{-k}(|v_1|) \tag{32}$$

and

$$|G[f_E](t,x+tv,v,v_1,\nu)| \leq \text{const.} W_{-p}(|x|)W_{1-k}(|v|)W_{1-k}(|v_1|) \qquad (33)$$

for all $t \geq 0$. Consequently condition (26) is satisfied.

Moreover for all fixed $(x,v,v_1,\nu) \in (R^3)^3 \times S^2$

$$t \longrightarrow G[f_E](t,x+tv,v,v_1,\nu)$$

is a continuous function on $[0, \infty)$ so that by the Lebesgue dominated convergence theorem it is possible to conclude that

$$P[f_E](t) = \int_{(R^3)^3 \times S^2} G[f_E](t,x+tv,v,v_1,\nu)d\nu dv dv_1 dx$$

is a continuous function on $[0,\infty)$, i. e.

$$P[f_E] \in C_b^o([0,\infty)). \qquad (34)$$

then

$$P^*[f_E] \in C^1([0,t]) \quad \text{for all} \quad t \in [0,\infty) \qquad (35)$$

Step 1 is, at this end, proven.

Step 2: Following Toscani [16], we can now choose a non-increasing sequence of functions

$$F_j(x,v) = \max\{F(x,v), (c_1/(1+j))W_{-p}(|x|)W_{-k}(|v|)\} \qquad (36)$$

and observe that

$$F_j \in C_b^o(R^3 x R^3), \qquad \lim_j |||F - F_j|||^o_{p,k} = 0. \qquad (37)$$

By Theorem 2, item ii), follows that there exists a sequence $\{f_{Ej}\} \in \overset{o}{\mathfrak{B}}_{p,k}$ of unique non-negative solutions f_{Ej} of problem (9) with initial data F_j in the place of F. Then, by Assumptions A.3-5 and using the same technique as in [16], it is possible to show that

$$|E^{\#}_{\mathfrak{G}}(f_E;f_E,f_E)(t,x,v)| \le \text{const. } W_{-p}(|x|)W_{2-k}(|v|) \tag{38}$$

Then, standard arguments, see f. i. [12], show that

$$f^{\#}_{Ej} \in C^1([0,\ t];\mathfrak{B}^{\circ}_{o,k-2}) \text{ for all } t \in [0, \infty) \tag{39}$$

and

$$df^{\#}_{Ej}/dt = E^{\#}_{\mathfrak{G}}(f_{Ej};f_{Ej},\ f_{Ej}) \tag{40}$$

Consequently

$$d(f^{\#}_{Ej}\ln f^{\#}_{Ej})/dt = (1+\ln f^{\#}_{Ej})df^{\#}_{Ej}/dt = (1+\ln f^{\#}_{Ej})E^{\#}_{\mathfrak{G}}(f_{Ej};f_{Ej},f_{Ej})$$

and

$$(1/\Delta)\{(f^{\#}_{Ej}\ln f^{\#}_{Ej})(t+\Delta) - (f^{\#}_{Ej}\ln f^{\#}_{Ej})(t)\} =$$

$$= (1/\Delta)\int_t^{t+\Delta}\{(1 + \ln f^{\#}_{Ej})E^{\#}_{\mathfrak{G}}(f_{Ej};f_{Ej},f_{Ej})\}(s)ds \tag{41}$$

The left hand side of equality (41) tends to $\partial(f^{\#}_{Ej}\ln f^{\#}_{Ej})/\partial t$ as $\Delta \to 0$ for all fixed $x \in R^3$ and $v \in R^3$. The absolute value of the right hand side of (41) can be estimated uniformly with respect to t and j by a function in $L_1(R^3 \times R^3)$ in the same fashion of ref.[16], see eqs.(18,19). Thus applying again the Lebesgue dominated convergence Theorem one can possible to conclude that

$$H[f_{Ej}] \in C^1([0,\ t]) \text{ for all } t \in [0,\infty) \tag{42}$$

and

$$H^*[f_{Ej}] \in C^1([0,\ t]) \text{ for all } t \in [0,\infty) \tag{43}$$

Moreover

$$dH^*[f_{Ej}]/dt = \int_{R^3 \times R^3} (1+\ln f_{Ej}) E_\sigma(f_{Ej}; f_{Ej}, f_{Ej}) dxdv - \tfrac{1}{2} P[f_{Ej}] \tag{44}$$

Using Polewczak's arguments [13] yields

$$\int_{R^3 \times R^3} ((1+\ln f_{Ej}) E_\sigma(f_{Ej}; f_{Ej}, f_{Ej}))(t,x,v) dxdv =$$

$$(1/2) \int_{(R^3)^3 \times S^2} Y^-(f_{Ej};\sigma)(t,x,\nu)\{f_{Ej}(t,x,v) f_{Ej}(t,x-\sigma\nu,v_1)$$

$$\ln(f_{Ej}(t,x,v') f_{Ej}(t,x-\sigma\nu,v_1')) - f_{Ej}(t,x,v) f_{Ej}(t,x-\sigma\nu,v_1)$$

$$\ln(f_{Ej}(t,x,v) f_{Ej}(t,x-\sigma\nu,v_1))\} \phi((v_1-v)\cdot\nu) d\nu dv_1 dxdv \tag{45}$$

Moreover noticing that $P[f]$ can be written in the form

$$P[f](s) = \int_{(R^3)^3 \times S^2} Y^-(f;\sigma)(s,x,\nu)\{f(s,x-\sigma\nu,v_1') f(s,x,v') -$$

$$- f(s,x-\sigma\nu,v_1) f(s,x,v)\} \phi((v_1-v)\ \nu) d\nu dv_1 dxdv$$

then the elementary inequality

$$a(\ln b - \ln a) \leq b - a \quad \text{for} \quad a > 0,\ b > 0 \tag{46}$$

where

$$a = f_{Ej}(t,x,v) f_{Ej}(t,x-\sigma\nu,v_1), \quad b = f_{Ej}(t,x,v') f_{Ej}(t,x-\sigma\nu,v_1')$$

can be applied to obtain

$$dH^*[f_{Ej}]/dt \leq 0 \qquad \text{for all } j \tag{47}$$

Thus, for all j and $0 \leq t_1 \leq t_2 < \infty$, we have

$$H^*[f_{Ej}](t_2) \leq H^*[f_{Ej}](t_1) \tag{48}$$

which proves the second step of the theorem.

Step 3: Using the same analysis of [16], see also [3], Chap.II and IV, shows that $\{f_{Ej}\}$ is a Cauchy sequence in $\overset{\circ}{\mathcal{B}}_{p,k}$ and its limit is the unique solution f_E of problem (9). Thus $(f^\#_{Ej} \ln f^\#_{Ej})(t,x,v)$ converges to $(f^\#_E \ln f^\#_E)(t,x,v)$ for all $t \in [0, \infty)$, $x \in R^3$ and $v \in R^3$. Moreover $G[f_{Ej}](t,x+vt,v,v_1,\nu)$ converges to $G[f_E](t,x+vt,v,v_1,\nu)$ for all $t \in [0,\infty)$, $x \in R^3$, $v \in R^3$, $v_1 \in R^3$ and $\nu \in S^2$.

Using (12) of ref.[16] shows that for all $t \in [0, \infty)$ the term $|f^\#_{Ej} \ln f^\#_{Ej}|(t)$ is uniformly bounded with respect to j by a function in $L_1(R^3 \times R^3)$. Thus by the Lebesgue dominated convergence theorem we obtain

$$\lim_{j \to \infty} H[f_{Ej}](t) = H[f_E](t) \quad \text{for all} \quad t \in [0,\infty) \tag{49}$$

moreover by (33)

$$\lim_{j \to \infty} P^*[f_{Ej}](t) = P^*[f_E](t) \quad \text{for all} \quad t \in [0,\infty) \tag{50}$$

then

$$\lim_{j \to \infty} H^*[f_{Ej}](t) = H^*[f_E](t) \quad \text{for all} \quad t \in [0,\infty) \tag{51}$$

therefore (27) is obtained by (48).

Taking into account Theorem 2 yields, with analogous arguments,

$$\lim_{\sigma \to 0} H[f_E](t) = H[f_B](t) \quad \text{for all} \quad t \in [0,\infty) \tag{52}$$

Moreover

$$\lim_{\sigma \to 0} P[f_E](t) = 0 \quad \text{for all} \quad t \in [0,\infty) \tag{53}$$

Thus (28) is satisfied and the Theorem is proven.■

Remark2: The main result has been obtained, as in [3] as well as in [8,9], if the Enskog equation is characterized by a radius of the hard spheres below a critical value σ. That has an immediate physical interpretation, $\sigma \leq \sigma_o$ implies that the assumptions leading to the Boltzmann equation can be reasonably extended to the Enskog equation. On the other hand, when $\sigma > \sigma_o$, this extension is certainly questionable.

Acknowledgments: This paper has been partially supported by the Ministery for Education and by the National Council for the Research under project MMAI of the National Group of Mathematical Physics. The second author has also been supported by the National Ministry for Education of Poland under grant n. RP.I.10 .

References

[1] Arkeryd L. and Cercignani C., To be published.
[2] Bellomo N. and Lachowicz M., Int. J. Modern Phys. B., 1, (1987), 1193-1206.
[3] Bellomo N. and Lachowicz M., J. Statist. Phys, 51, (1988), 233-247.
[4] Bellomo N., Palczewski A. and Toscani G., Mathematical Topics in Nonlinear Kinetic Theory, World Scientific, London, Singapore, (1988).
[5] Di Perna R. and Lions P.L, Comp. Rend. Acad. Sci. Paris, t.306, I, (1988), 343-346.
[6] Di Perna R. and Lions P.L., Internal Report, Univ. de Paris Dauphine, n. 8903, (1989).
[7] Enskog D. Svenska Akad., 4, (1921), 3-44.
[8] Lachowicz M., Arch. Rational. Mech. Anal., 101, (1988), 179-194.
[9] Lachowicz M., Ann.Mat. Pura Appl., To appear.
[10] Lebowitz J., Percus J. and Sykes J., Physical Review, 188, (1968), 487-507.
[11] Marechal M., Blawzdziewicz J. and Piasecki J., Phys. Rew. Lett., 52, (1984), 1169-1174.
[12] Polewczak J., SIAM J. Appl. Math., To appear.
[13] Polewczak J., J. Statist. Phys, To Appear.

[14] Resibois P. and De Leener M., Classical Kinetic Theory of Fluids, Wiley, London, (1977).

[15] Resibois P., J. Stat. Phys., 19, (1978), 593-609.

[16] Toscani G., Arch. Rational. Mech. Anal.,100, (1987), 1-12.

[17] Van Beijeren H. and Ernst M. H., Physica, 68, (1973), 437-456.

[18] Van Beijeren H., Phys. Rew. Lett., 51, (1983), 1503-1506.

On some numerical problems in semiconductor device simulation *

F. Brezzi[1][2]-*L.D. Marini*[2]-*P. Markowich*[3]- *P. Pietra*[2]

[1] Istituto di Meccanica Strutturale, Università di Pavia(Italy)
[2] Istituto di Analisi Numerica del C.N.R. di Pavia(Italy)
[3] Fachbereich Mathematik, TU-Berlin (FRG)

ABSTRACT - We recall in the introduction the main features of the drift-diffusion model for semiconductor devices, pointing out its physical meaning, its possible derivation, and its limits. Then, in Section 2, we present a mixed finite element method for the discretization of this model. Finally, using asymptotic analysis techniques, we compare the qualitative behaviour of the mixed method with other methods (classical conforming Galerking method and harmonic average methods). This asymptotic analysis provides some indication of the advantages of the mixed method.

1. Introduction

The most commonly used model for charge transport in semiconductors is the so called drift-diffusion model, which - in an appropriate system of units - reads

$$\frac{\partial p}{\partial t} + div_x \underline{J} = -R \tag{1.1}$$

$$\underline{J} = -\mu(\underline{\nabla}_x p + p\underline{\nabla}_x \psi) \,. \tag{1.2}$$

Here, p denotes the position density of the positively charged holes, $\underline{J}$ the hole current density, and ψ the electrostatic potential. $\underline{E} = -\underline{\nabla}_x \psi$ is the electric field, and the source term R is the recombination-generation rate of charged carrier pairs. The coefficient $\mu > 0$ stands for the hole mobility. The equations (1.1), (1.2) hold for all values of the position variable x in the semiconductor domain $\Omega \subset \mathbf{R}^3$, and for the time $t \geq 0$ (in the sequel we shall skip the subscript x when denoting the gradient and the divergence taken with respect to the position variable).

The drift-diffusion equations are supplemented by mixed Neumann Dirichlet boundary conditions for p on $\partial\Omega$, and by an initial condition at $t = 0$. The Dirichlet segments

* Partially supported by C.N.R. Sp. proj. on Informatic systems and Parallel comput. and C.N.R. contr. 88.00326.01

model contacts, where a voltage is applied to the device, and the Neumann segments represent insulating boundaries or so called artificial boundary segments introduced to separate the device under consideration from neighbouring devices in the chip. For details on the boundary and initial conditions we refer to [9], [10].

The corresponding drift-diffusion model for the position density n of the negatively charged conduction electrons is obtained by making the obvious sign changes in (1.1),(1.2), taking into account the opposite flow direction of the electrons in the electric field $\underline{E}$. The potential ψ is usually modelled self-consistently as Coulomb potential generated by the space charge of the semiconductor, i.e., it is obtained as solution of a Poisson equation whose right-hand side represents the charge density determined by p, n, and by the density of impurity ions implanted into the semiconductor in order to control its electrical performance (see [16]). Then, after modelling the recombination-generation rate and the carrier mobilities as functions of n, p, and $\underline{E}$ and after inserting the current relation (1.2) into the continuity equation (1.1), a system of two parabolic equations (for n and p respectively) coupled to the Poisson equation for ψ is obtained. The mathematical analysis and the numerical treatment of this highly non linear pde system is a formidable task and has received a lot of attention in the mathematical, physical, and engineering literature so far (see [9], [10], [16] and the references therein). In this paper we shall focus on a particular issue, namely on the discretization of the steady-state drift-diffusion equation in the two dimensional case, i.e., we shall assume $\frac{\partial p}{\partial t} \equiv 0$, and $\Omega \subset \mathbf{R}^2$. Before going into the details of the numerical analysis (Sections 2 and 3), we now discuss the physical background of the drift-diffusion model.

Obviously, the equation (1.1) is a standard conservation law (continuity equation) for the hole current density, analogous to fluid dynamics. The source term R describes the loss and, respectively, gain of charged particles due to the recombination-generation of electron-hole carrier pairs (see [16] for an account of the most important recombination-generation mechanisms in semiconductor).

The current relation (1.2) also has a simple phenomenological interpretation. Obviously, the term $\underline{J}_{diff} := -\mu \underline{\nabla} p$ represents a diffusion current, and $\underline{J}_{drift} := \mu p \underline{E}$ a drift current caused by the electric field $\underline{E}$. Thus, (1.2) postulates that the total hole current in the semiconductor is the sum of a diffusive current and of an electric field driven convection current. This intriguing but purely phenomenological reasoning however does not explain the equality of the diffusion and convection coefficients. Moreover, a more rigorous derivation which illuminates the limits of validity of the drift-diffusion model is desirable. Both these issues are dealt with in a satisfactory way by taking the semiclassical Boltzmann equation for semiconductors (see [10], [13]) as a basis for deriving (1.1),(1.2). The Boltzmann equation, whose solution is the phase space density of holes, models the convection of the electric field by a hyperbolic differential operator in the phase space and the collisions (scattering events) of particles with each other and with their environment by a non local non linear operator. It turns out that the drift-diffusion model can be derived from the Boltzmann equation by taking the limit of the normed mean free path (i.e., the average length travelled by a particle between two consecutive scattering events divided by the characteristic length of the device) going to zero, when a linearized scattering operator is used (see [10], [14]). The equality of the diffusion and drift coefficients comes out automatically in this approach. Also, information on the validity of (1.1),(1.2) can be deduced from this limiting procedure. First, the considered physical situation must be such that the performed linearization of the scattering kernel is valid, i.e., the particle density must be reasonably small. This requires a non-degenerate semiconductor (the density of the implanted impurity ions is reasonably low, i.e., at most $\simeq 10^{19}$ particles per cubic centimeter in silicon). Also, the semiconductor has to be operated under moderate electric field strengths. Second, the normed mean free path of the device must be small in order for the limiting procedure to make sense. This means that the caracteristic length of the considered device must be large in comparison with the mean free path of the semiconductor. While this condition is usually satisfied for modern silicon technology, it fails for other semiconductors which have a larger mean free path. However, it is certainly also going to be a problem for the next generation of even higher integrated silicon devices (technology with the characteristic device length of less than 0.5 micron).

We remark that a lot of research on other charge transport models for semiconductors is currently going on. Quantum transport models (Wigner equation) are investigated for ultra integrated semiconductor devices, semiclassical Boltzmann-type models for semiconductors other than silicon and, most recently,the so called hydrodynamic semiconductor transport model,which is an extension of the drift-diffusion model not based on the mean free path limit, is under extensive scrutiny (see [10]).

2. Mixed approximation of the continuity equation

We shall describe in this section a mixed approximation to the continuity equation (1.1), (1.2). For simplicity, we shall consider the stationary two-dimensional case and a constant mobility coefficient $\mu \equiv 1$. In the solution of the coupled system of three equations (for ψ, n, and p), a linearization method of Gummel type (approximate Newton decoupling method, [9], [16]) is often used. Then, at each iteration, one has to solve a problem of the type

$$\begin{cases} \textit{Find } p \in H^1(\Omega) \textit{ such that} \\ -div(\underline{\nabla} p + p\underline{\nabla}\psi) \; + \; cp \; = \; f \quad \textit{in } \Omega \subset \mathbf{R}^2 \\ \qquad p \; = \; g \quad \textit{on } \Gamma_0 \subset \partial\Omega \\ \qquad \dfrac{\partial p}{\partial n} \; = \; 0 \quad \textit{on } \Gamma_1 = \partial\Omega \backslash \Gamma_0 \end{cases} \tag{2.1}$$

where ψ is assumed to be known and piecewise linear (coming from a discretization of the Poisson equation). In the equation (2.1) f is a function independent of p, and c a non negative function independent of p, which can be assumed piecewise constant. To simplify the exposition, we shall assume here $c = 0$. We refer to [8] for the treatment of the more general case $c \geq 0$. We recall that, since $|\underline{\nabla}\psi|$ is quite large in some parts of the domain, equation (2.1) is an advection dominated equation, for which classical discretization methods may fail. Using the classical change of variable from the charge density p to the Slotboom variable ρ

$$p \; = \; \rho e^{-\psi} \, , \tag{2.2}$$

equation (2.1) can be written in the symmetric form

$$\begin{cases} \textit{Find } \rho \in H^1(\Omega) \textit{ such that} \\ \qquad -div(e^{-\psi}\underline{\nabla}\rho) \; = \; f \qquad \textit{in } \Omega \\ \qquad \rho \; = \; \chi \; := \; e^{\psi} g \quad \textit{on } \Gamma_0 \\ \qquad \dfrac{\partial \rho}{\partial n} \; = \; 0 \qquad \textit{on } \Gamma_1 \end{cases} \tag{2.3}$$

and the hole current density is now given by

$$\underline{J} \; = \; -e^{-\psi}\underline{\nabla}\rho \, . \tag{2.4}$$

Note that in (2.3) homogeneous Neumann conditions come from the usually made assumption that $-\underline{E} \cdot \underline{n} \equiv \frac{\partial \psi}{\partial n}$ vanishes on Γ_1. The idea is to discretize equation (2.3) with mixed finite element methods, go back to the original variable p by using a discrete version of the transformation (2.2), and then solve for p. For the case $c = 0$, a mixed scheme (based on the lowest order Raviart-Thomas element [15]) has been introduced and extensively discussed in [4] for the case $f = 0$, and in [5] for $f \neq 0$. The scheme provides an approximate current with continuous normal component at the interelement boundaries. Moreover, the matrix associated with the scheme can be proved to be an M-matrix, if a weakly acute triangulation is used (every angle of every triangle is $\leq \pi/2$). This property guarantees a discrete maximum principle and, in particular, a non-negative solution if the boundary data are non-negative. Moreover, when going back to the variable p, this structure property of the matrix is retained.

Let us recall the mixed scheme. For that, let $\{T_h\}$ be a regular decomposition of Ω into triangles T ([6]) (Ω is assumed to be a polygonal domain). According to [15], we define, for all T$\in T_h$, the following set of polynomial vectors

$$RT(\mathrm{T}) \;=\; \{\underline{\tau} = (\tau_1, \tau_2),\ \tau_1 = \alpha + \beta x,\ \tau_2 = \gamma + \beta y,\ \alpha, \beta, \gamma \in \mathbf{R}\}\ . \tag{2.5}$$

Then, we construct our finite element spaces as follows

$$\widetilde{V}_h = \{\underline{\tau} \in [L^2(\Omega)]^2 : \ div\underline{\tau} \in L^2(\Omega),\ \underline{\tau} \cdot \underline{n} = 0 \ on\ \Gamma_1,\ \underline{\tau}_{|\mathrm{T}} \in RT(\mathrm{T}),\ \forall \mathrm{T} \in T_h\}, \tag{2.6}$$

$$W_h = \{\phi \in L^2(\Omega) : \phi_{|\mathrm{T}} \in P_0(\mathrm{T})\ \forall \mathrm{T} \in T_h\}. \tag{2.7}$$

As usual, $P_0(\mathrm{T})$ denotes the space of constants on T. The mixed discretization of (2.3) is then the following

$$\begin{cases} Find\ \underline{\widetilde{J}}_h \in \widetilde{V}_h\ ,\ \widetilde{\rho}_h \in W_h\ such\ that: \\ \displaystyle\int_\Omega e^{\overline{\psi}} \underline{\widetilde{J}}_h \cdot \underline{\tau} dxdy - \int_\Omega div\ \underline{\tau}\ \widetilde{\rho}_h\ dxdy = 0 \quad \underline{\tau} \in \widetilde{V}_h, \\ \displaystyle\int_\Omega div\ \underline{\widetilde{J}}_h\ \phi\ dxdy = \int_\Omega f\phi dxdy \qquad \phi \in W_h. \end{cases} \tag{2.8}$$

In the first equation of (2.8) $\overline{\psi}$ denotes the piecewise constant function defined in each triangle T by

$$e^{\overline{\psi}}{}_{|\mathrm{T}} \;=\; (\int_\mathrm{T} e^{\psi} dxdy)/|\mathrm{T}|\ . \tag{2.9}$$

It is clear that $\widetilde{\rho}_h$ will be an approximation of the solution ρ of (2.3), and $\widetilde{\underline{J}}_h$ will be an approximation of the current $\underline{J}$. In particular, the first equation of (2.8) is a discretized version of (2.4), and the second equation of (2.8) is a discretized version of $div\underline{J} = f$. Uniqueness results for (2.8) follow from the general theory of [3].

We remark that the condition $div\underline{\tau} \in L^2(\Omega)$ in the definition (2.6) implies that *every* $\underline{\tau} \in \widetilde{V}_h$ *has a continuous normal component* when going from one element to another. This means, in particular, that the current is preserved.

The algebraic treatment of system (2.8) needs some care. Actually, the matrix associated with (2.8) has the form

$$\begin{pmatrix} \widetilde{A} & -\widetilde{B} \\ -\widetilde{B}^* & 0 \end{pmatrix} \tag{2.10}$$

and is not positive-definite (H^* denotes the transpose of the matrix H). A way to avoid this inconvenience is to relax the continuity requirement in the space definition (2.6) and to enforce it back by using interelement Lagrange multipliers. (See [7] where this idea was first introduced). The procedure is the following. First we set

$$V_h \;=\; \{\underline{\tau} \in [L^2(\Omega)]^2 : \underline{\tau}_{|\mathrm{T}} \in RT(\mathrm{T})\ \forall \mathrm{T} \in T_h\}. \tag{2.11}$$

Then, denoting by E_h the set of edges e of T_h, we define, for any function $\xi \in L^2(\Gamma_0)$

$$\Lambda_{h,\xi} \;=\; \{\mu \in L^2(E_h) : \mu_{|e} \in P_0(e)\ \forall e \in E_h\ ;\ \int_e (\mu - \xi)ds = 0\ \forall e \subset \Gamma_0\}\ , \tag{2.12}$$

where $P_0(e)$ denotes the space of constants on e. The mixed-equilibrium discretization of (2.3) is then

$$\begin{cases} \textit{Find } \underline{J}_h \in V_h\ ,\ \rho_h \in W_h\ ,\ \lambda_h \in \Lambda_{h,\chi} \textit{ such that}: \\ \displaystyle\int_\Omega e^{\overline{\psi}} \underline{J}_h \cdot \underline{\tau} dxdy - \sum_{\mathrm{T}} \int_{\mathrm{T}} div\ \underline{\tau}\ \rho_h dxdy + \sum_{\mathrm{T}} \int_{\partial \mathrm{T}} \lambda_h \underline{\tau} \cdot \underline{n}\ ds = 0 & \underline{\tau} \in V_h, \\ \displaystyle\sum_{\mathrm{T}} \int_{\mathrm{T}} div\ \underline{J}_h\ \phi dxdy = \int_\Omega f \phi dxdy & \phi \in W_h, \\ \displaystyle\sum_{\mathrm{T}} \int_{\partial \mathrm{T}} \mu \underline{J}_h \cdot \underline{n}\ ds \;=\; 0 & \mu \in \Lambda_{h,0}. \end{cases} \tag{2.13}$$

It is easy to see that problem (2.13) has a unique solution and that

$$\underline{J}_h \;\equiv\; \widetilde{\underline{J}}_h\ ,\ \rho_h \;\equiv\; \widetilde{\rho}_h\ . \tag{2.14}$$

Moreover, λ_h is a good approximation of ρ at the interelements. (See [1] for detailed proofs). The linear system associated with (2.13) can be written in matrix form as

$$\begin{pmatrix} A & -B & C \\ -B^* & 0 & 0 \\ C^* & 0 & 0 \end{pmatrix} \begin{pmatrix} \underline{J}_h \\ \rho_h \\ \lambda_h \end{pmatrix} = \begin{pmatrix} 0 \\ -F \\ 0 \end{pmatrix} . \tag{2.15}$$

In (2.15) the notation $\underline{J}_h, \rho_h, \lambda_h$ is used also for the vectors of the nodal values of the corresponding functions. The matrix in (2.15) is not positive definite. However, A is block- diagonal (each block being a 3x3 matrix corresponding to a single element T) and can be easily inverted at the element level. Hence, the variable $\underline{J}_h$ can be eliminated by static condensation, leading to the new system

$$\begin{pmatrix} B^*A^{-1}B & -B^*A^{-1}C \\ & \\ -C^*A^{-1}B & C^*A^{-1}C \end{pmatrix} \begin{pmatrix} \rho_h \\ \\ \lambda_h \end{pmatrix} = \begin{pmatrix} F \\ \\ 0 \end{pmatrix} . \tag{2.16}$$

The matrix in (2.16) is symmetric and positive definite. Moreover, $B^*A^{-1}B$ is a diagonal matrix, so that the variable ρ_h can also be eliminated by static condensation. This leads to a final system, acting on the unknown λ_h only, of the form

$$M\,\lambda_h = G\ , \tag{2.17}$$

where M and G are given by:

$$M = C^*A^{-1}C - C^*A^{-1}B(B^*A^{-1}B)^{-1}B^*A^{-1}C\ , \tag{2.18}$$

$$G = C^*A^{-1}B(B^*A^{-1}B)^{-1}F\ , \tag{2.19}$$

and M is symmetric and positive definite. In order to go back to the original unknown p we recall that λ_h is an approximation of ρ and we can use a discrete version of the inverse transform of (2.2):

$$\lambda_h = (\mathrm{e}^{\psi})^I p_h. \tag{2.20}$$

In (2.20) $(\mathrm{e}^{\psi})^I$ is given edge by edge by the meanvalue of e^{ψ}:

$$\mathrm{e}^{\psi^I}\big|_e = \Big(\int_e \mathrm{e}^{\psi} ds\Big)/|e|\ . \tag{2.21}$$

The transformation (2.20) amounts to multiplying the matrix M columnwise by the value of $(\mathrm{e}^{\psi})^I$ on the corresponding edge. The final system in the unknown p_h will be of the type

$$\widetilde{M} p_h = G\ . \tag{2.22}$$

The matrix $\widetilde{M}$ is not symmetric anymore, but it is an M-matrix if the matrix (2.18) is an M-matrix, which holds true if the triangulation is of weakly acute type.

3. Asymptotic behaviour of the numerical scheme

We already pointed out in the previous sections that the electric field $\underline{E}$ $(= -\underline{\nabla}\psi)$ can be, in most applications, very large in some parts of the domain Ω. The aim of this section is to perform a (rough) analysis of the mixed exponential fitting scheme (and of some other possible schemes for (2.3)) when the electric field becomes larger and larger. This will show why the choice of a mixed method for discretizing (2.3) seems to be preferable, apart from the obvious reason that it is strongly current-preserving.

In order to perform our asymptotic analysis we shall make the simplifying assumption that we are dealing with a given potential ψ, piecewise linear, of "moderate size", and that our equation is

$$-div(\underline{\nabla}p + p\underline{\nabla}(\frac{\psi}{\lambda})) = f, \tag{3.1}$$

where λ is a real valued parameter. We are obviously interested in the behaviour of numerical schemes for (3.1) when λ becomes smaller and smaller. The symmetric form of (3.1) reads then

$$-div(e^{-(\psi/\lambda)}\underline{\nabla}\rho) = f, \tag{3.2}$$

where the change of variable is now

$$p = \rho e^{-(\psi/\lambda)} . \tag{3.3}$$

We shall analyze the asymptotic behaviour (as $\lambda \to 0$) of three different schemes, all based on the idea of discretizing (3.2) first, and then use (3.3) to obtain a numerical scheme in the unknown p (and hence a scheme for (3.1)). In particular, we will consider the following discretization methods for (3.2): *a)* classical conforming piecewise linear methods, *b)* conforming piecewise linear mothods with harmonic average (as pointed out in [4], [5] they can be regarded as a discretization of (3.2) by means of hybrid methods), and *c)* mixed methods as described in the previous section.

We recall that, calling Z_h the space of continuous piecewise linear functions on Ω and setting, for all function $\xi \in C^0(\overline{\Gamma}_D)$

$$Z_{h,\xi} = \{v \in Z_h,\ v = \xi \ at\ nodes\ \in \overline{\Gamma}_D\}\ , \tag{3.4}$$

the methods *a)* and *b)* can be written in the following way.

Classical method

$$\begin{cases} (i) & \rho_h \in Z_{h,\chi} , \\ (ii) & \int_\Omega e^{-(\psi/\lambda)} \underline{\nabla}\rho_h \cdot \underline{\nabla} v dxdy = \int_\Omega f v dxdy \quad \forall v \in Z_{h,0} , \\ (iii) & p_h = e^{-(\psi/\lambda)} \rho_h \quad \textit{at the nodes} . \end{cases} \tag{3.5}$$

Conforming method with harmonic average

$$\begin{cases} (i) & \rho_h \in Z_{h,\chi} , \\ (ii) & \int_\Omega \overline{e^{-(\psi/\lambda)}} \underline{\nabla}\rho_h \cdot \underline{\nabla} v dxdy = \int_\Omega f v dxdy \quad \forall v \in Z_{h,0} , \\ (iii) & p_h = e^{-(\psi/\lambda)} \rho_h \quad \textit{at the nodes} , \\ (iv) & \overline{e^{-(\psi/\lambda)}}_{|T} = \dfrac{|T|}{(\int_T e^{\psi/\lambda} dxdy)} \quad \forall T \in T_h \ (\textit{harmonic average}). \end{cases} \tag{3.6}$$

In order to analyze the behaviour of the schemes (3.5), (3.6) and of the mixed scheme of Section 2, we shall need the following asymptotic formulae, valid as $\lambda \to 0$ for a function ϕ linear on a triangle T:

$$\int_T e^{\phi/\lambda} dxdy \simeq \lambda^2 |T| e^{\phi^T_{max}/\lambda}, \tag{3.7}$$

$$\int_e e^{\phi/\lambda} ds \simeq \lambda |e| e^{\phi^e_{max}/\lambda}, \tag{3.8}$$

In (3.7), (3.8) T is a triangle, e is an edge of T, and ϕ^T_{max}, ϕ^e_{max} represent the maximum value of ϕ over $\overline{T}$ and over $\overline{e}$, respectively. Formulae (3.7), (3.8) can be easily checked by direct computation. They hold in the *generic* case where the values ϕ^T_{max} and ϕ^e_{max} are assumed only at one point.

We are now able to analyze the limit behaviour of the various schemes. For this, let us just look at the contributions of a single triangle T to the final matrix. Denoting by $\phi^{(i)}$ $(i = 1, 2, 3)$ the basis functions on T, we have, for the classical method

$$\int_T e^{-(\psi/\lambda)} \underline{\nabla}\phi^{(i)} \cdot \underline{\nabla}\phi^{(j)} dxdy \simeq \lambda^2 L^T_{ij} e^{-(\phi^T_{min}/\lambda)}, \tag{3.9}$$

where L^T_{ij} are the contributions of the conforming approximation of the Laplace operator, that is

$$L^T_{ij} := \int_T \underline{\nabla}\phi^{(i)} \cdot \underline{\nabla}\phi^{(j)} dxdy, \tag{3.10}$$

and $\psi^{\rm T}_{min}$ is obviously the minimum value of ψ over $\overline{\rm T}$. Taking into account transformation (3.5,*iii*), the contributions to the final matrix, acting on p_h are given by

$$M^{\rm T}_{ij} \simeq \lambda^2 L^{\rm T}_{ij} \mathrm{e}^{(\psi_j - \psi^{\rm T}_{min})/\lambda}, \tag{3.11}$$

(where ψ_j= value of ψ at the node j). Hence, for the classical method, some coefficients of the matrix blow up exponentially when $\lambda \to 0$.

Let us now consider the case (3.6) where the harmonic average is used. From (3.6,*iv*) and (3.7) we have

$$\int_{\rm T} \overline{\mathrm{e}^{-(\psi/\lambda)}} \underline{\nabla}\phi^{(i)} \cdot \underline{\nabla}\phi^{(j)} \, dxdy \simeq \frac{1}{\lambda^2} L^{\rm T}_{ij} \mathrm{e}^{-(\psi^{\rm T}_{max})/\lambda}, \tag{3.12}$$

where the coefficients $L^{\rm T}_{ij}$ are defined in (3.10). Then, combining (3.12) and (3.6 ,*iii*), the contributions of the triangle T to the final matrix are

$$M^{\rm T}_{ij} \simeq \frac{1}{\lambda^2} L^{\rm T}_{ij} \mathrm{e}^{(\psi_j - \psi^{\rm T}_{max})/\lambda}. \tag{3.13}$$

We see that, when using the harmonic average, some contributions can become very small, but this can be regarded as a natural upwinding effect which is rather desirable than disturbing. However, it is also clear that the contributions which are not exponentially small have order of magnitude $1/\lambda^2$, while from (3.1) one would expect coefficients of order $1/\lambda$. As discussed in [5] in the framework of hybrid methods, this is clearly not disturbing if $f = 0$, but it can be a source of inconsistency for $f \neq 0$ and λ small, as shown in [5] on simple practical experiments. (We refer to [5] for possible remedies for this method). We point out that this drawback is not present in the mixed formutation (2.13), (2.9), (2.21). Actually, one can easily see that the contributions of a triangle T to the final matrix, acting on p_h, for mixed methods are given by

$$M^{\rm T}_{ij} = \int_{\rm T} \overline{\mathrm{e}^{-(\psi/\lambda)}} \underline{\nabla}\chi^{(i)} \cdot \underline{\nabla}\chi^{(j)} \, dxdy \, \Big(\int_{e_j} \mathrm{e}^{\psi/\lambda} ds\Big) \, |e_j|^{-1}, \tag{3.14}$$

where the harmonic average (3.6,*iv*) is used. In (3.14) e_i ($i = 1,2,3$) are the edges of T, and $\chi^{(i)}$ are the piecewise linear non-conforming basis functions, that is,

$$\chi^{(i)} \in P_1(\mathrm{T}) \; ; \; \int_{e_j} \chi^{(i)} ds = |e_j| \delta_{ij}, \tag{3.15}$$

where δ_{ij} is the Kronecker's symbol. From (3.14), (3.6,*iv*), (3.7), and (3.8) we have then

$$M_{ij}^{\mathrm{T}} \simeq \frac{1}{\lambda}\widetilde{L}_{ij}^{\mathrm{T}} e^{(\psi_{max}^{e_j}-\psi_{max}^{\mathrm{T}})/\lambda}, \tag{3.16}$$

where $\widetilde{L}_{ij}^{\mathrm{T}}$ are the coefficients of the elementary stiffness matrix coming from a piecewise linear non-conforming approximation of the Laplace operator, that is,

$$\widetilde{L}_{ij}^{\mathrm{T}} := \int_{\mathrm{T}} \underline{\nabla}\chi^{(i)} \cdot \underline{\nabla}\chi^{(j)}\, dxdy. \tag{3.17}$$

It is now clear what the advantages of mixed methods are: 1) exponential blow-up of the coefficients is avoided, 2) some contributions will go exponentially to zero, corresponding to a natural upwinding effect, 3) the order of magnitude of the non vanishing coefficients is $1/\lambda$, as expected from (3.1).

The above considerations shed, in our opinion, a better light on several common choices for finite element approximations of the continuity equations, motivating the use of ***one-dimensional harmonic averages*** which are common in semiconductor device applications ([2], [12], [11] etc.). In the context of mixed methods we can use two-dimensional harmonic averages (which is, in a sense, more natural), since we compensate a factor λ from (3.8), due to the different change of variable from ρ_h to p_h (average on an edge instead of point value).

Remark We discussed so far the generic case where $\psi_{|\mathrm{T}}$ reaches its maximum at one point only. However, one can easily see that the "automatic adjustment" provided by mixed methods works as well for the non generic case where $\psi_{|\mathrm{T}}$ reaches its maximum on a whole edge. Finally, for $\psi = constant$ on T, we are just dealing with the Laplace operator, for which usual and harmonic average coincide and both give rise to the standard conforming scheme for Laplace operator. Similarly, the mixed approach above described produces the usual mixed approximation of the Laplace operator.

4. References

[1] D.N.Arnold - F.Brezzi: Mixed and non-conforming finite element methods: implementation, post-processing and error estimates. M^2AN **19**, 7-32, 1985.

[2] R.E.Bank - D.J.Rose - W.Fichtner: Numerical methods for semiconductor device simulation. *IEEE Trans. El. Dev.* **30**, 1031-1041, 1983.

[3] F.Brezzi: On the existence uniqueness and approximation of saddle-point problems arising from Lagrangian multipliers. *R.A.I.R.O.* **8-R2**, 129-151, 1974.

[4] F.Brezzi - L.D. Marini - P.Pietra: Two-dimensional exponential fitting and applications to drift-diffusion models. (To appear in SIAM J.Numer.Anal.).

[5] F.Brezzi - L.D. Marini - P.Pietra: Numerical simulation of semiconductor devices. (To appear in Comp.Meths.Appl. Mech.and Engr.).

[6] P.G.Ciarlet: *The Finite Element Method for Elliptic Problems.* North-Holland, Amsterdam, 1978.

[7] B.X.Fraeijs de Veubeke: Displacement and equilibrium models in the finite element method. In: *Stress Analysis*, O.C.Zienkiewicz and G.Hollister eds., Wiley, New York, 1965.

[8] L.D.Marini - P.Pietra: New mixed finite element schemes for current continuity equations. (Submitted to COMPEL).

[9] P.A.Markowich: *The Stationary Semiconductor Device Equations.* Springer, 1986.

[10] P.A.Markowich - C.Ringhofer - C.Schmeiser: *Semiconductor equations.* Springer, 1989. (To appear).

[11] P.A.Markowich - M.Zlámal: Inverse-average-type finite element discretisations of self-adjoint second order elliptic problems, *Math. of Comp.* **51**, 431-449, 1988.

[12] M.S.Mock: Analysis of a discretisation algorithm for stationary continuity equations in semiconductor device models II. *COMPEL* **3**, 137-149, 1984.

[13] B.Niclot - P.Degond - F.Poupaud: Deterministic particle simulations of the Boltzmann transport equation of semiconductors. J. Comp. Phys., **78**, 313-350, 1988.

[14] F.Poupaud: On a system of nonlinear Boltzmann equations of semiconductor physics. (To appear in SIAM J. Math. Anal.).

[15] P.A.Raviart - J.M.Thomas: A mixed finite element method for second order elliptic problems. In *Mathematical aspects of the finite element method*, Lecture Notes in Math. **606**, 292-315, Springer, 1977.

[16] S.Selberherr: *Analysis and simulation of semiconductor devices.* Springer, 1984.

CANONICAL PROPAGATORS FOR NONLINEAR SYSTEMS: THEORY AND SAMPLE APPLICATIONS

Dan G. Cacuci
Department of Chemical and Nuclear Engineering
University of California, Santa Barbara
Santa Barbara, CA 93106

and

V. Protopopescu
Engineering Physics and Mathematics Division
Oak Ridge National Laboratory
Oak Ridge, TN 37831-6363

ABSTRACT

A new canonical formalism for solving general nonlinear systems is presented. Fundamental to this formalism is the construction of forward (advanced) and backward (retarded) propagators that yield the problem's solution *exactly*, by propagating volume, surface, and initial sources. These propagators also satisfy reciprocity and semigroup properties. Therefore, they represent a generalization to nonlinear systems of the Green's functions from linear theory. Several examples are presented to illustrate the application of the algorithm as well as its numerical advantages over alternative methods.

1. INTRODUCTION

One of the most powerful methods for solving linear initial boundary value problems relies on Green's function formalism. Within this framework, the solution u of a linear system $(Lu)(x) = f(x)$ is given by $u(x) = \int f(x')G^*(x',x)dx' = \int f(x')G(x,x')dx'$, where the forward (advanced) and backward (retarded) Green's functions $G^*(x,x')$, $G(x,x')$ are solutions of the equations $(L^*G^*)(x,x') = \delta(x-x')$ and $(LG)(x,x') = \delta(x-x')$, respectively.

Recently, a new formalism has been proposed[1,2] to generalize the Green's function formalism for nonlinear systems. The formalism exploits exactly the methods of linear theory without resorting (at least in principle) to any approximations. Specifically, we construct forward (advanced) and backward (retarded) propagators which generalize the customary Green's functions and which reduce exactly to them for linear problems.

The construction of the propagators for nonlinear systems is presented in Section 2. The solution of the nonlinear system is then expressed as an integral containing the propagator as the respective kernel. Closed-form integral equations satisfied by the propagators are derived in Section 3. Several illustrative examples and computational aspects are presented in Section 4. The conclusions of our work are summarized in Section 5.

2. CONSTRUCTION OF PROPAGATORS FOR NONLINEAR SYSTEMS

A general nonlinear systems can be represented in abstract form as

$$N(u) = f \tag{2.1}$$

subject to boundary and/or initial conditions

$$\Gamma(u) = g\,, \tag{2.2}$$

where $u = (u_1, \ldots, u_n)$ is an n-component vector describing the state (i.e., dependent) variables, $N = (N_1, \ldots, N_n)$ is an n-component vector representing specific nonlinear (differential, integral, algebraic) operators; $f = (f_1, \ldots, f_n)$ denotes the volume sources; $\Gamma = (\Gamma_1, \ldots, \Gamma_j)$ represents the boundary operators, and $g = (g_1, \ldots, g_j)$ are the boundary and/or initial sources. Throughout the paper we shall be concerned only with the algorithmic and applicative aspects of the formalism. Since existence, uniqueness, and regularity issues could be tackled only within an explicit and problem-adapted functional framework, all the operations to follow are to be considered formal. At this stage, the justification of the formalism is based only on an *a posteriori* examination and on numerical performance. For illustration purposes, we consider $u(x)$, $x \in \Omega \subset \mathbf{R}^m$, to be an element in the real Hilbert space $L_2(\Omega) \times \ldots \times L_2(\Omega) = L_2^n(\Omega)$ endowed with the inner product

$$< u, v > \equiv \sum_{i=1}^{n} \int_{\Omega} u_i(x) v_i(x) dx\,. \tag{2.3}$$

It is convenient to combine Eqs. (2.1) and (2.2) into the single expression

$$N(u) + \delta \cdot \Gamma(u) = f + \delta \cdot g \tag{2.4}$$

by introducing symbolic δ-distributions associated in a unique manner with the direct boundary spaces of the problem (2.1) and (2.2) (see, *e.g.*, Ref. 3). In the formulation (2.4), the variables x will take values in the closure of Ω that we shall denote by $\bar{\Omega}$. Using this representation, we can consider without loss of generality that $N(0) = 0$ and $\Gamma(0) = 0$, since all inhomogeneities can be incorporated in f and g.

To construct the propagators for the nonlinear system represented by Eq. (2.4), we require that the first Gâteaux-derivatives of the operators appearing on the left side of Eq. (2.4) exist. These Gâteaux-derivatives are defined as

$$[N'(u) + \delta \cdot \Gamma'(u)]\, h \equiv \{d/d\epsilon\, [N(u + \epsilon h) + \delta \cdot \Gamma(u + \epsilon h)]\}_{\epsilon=0}\,; \tag{2.5}$$

the operator $N'(u) + \delta \cdot \Gamma'(u)$ represents an $n \times n$ matrix whose (ij)-th element is the operator $\partial[N_i(u) + \delta \cdot \Gamma_i(u)]/\partial u_j$. Note that the operator $N'(u) + \delta \cdot \Gamma'(u)$ depends nonlinearly on the parameter u but acts linearly on the n-component vector h; h is a fixed (but otherwise arbitrary) vector in $L_2^n(\Omega)$.

Since the operator $N'(u) + \delta \cdot \Gamma'(u)$ acts linearly on h, its adjoint $N'^* + \delta^* \cdot \Gamma'^*$ can be introduced via the customary linear duality, i.e.

$$< [N'(u) + \delta \cdot \Gamma'(u)]h, v > = < h, [N'^*(u) + \delta^* \cdot \Gamma'^*(u)]v > . \tag{2.6}$$

In Eq. (2.6), v is an arbitrary n-component vector in the dual space; in our case, the dual space is $L_2(\Omega) \times \ldots \times L_2(\Omega) = L_2^n(\Omega)$ since L_2 is self-dual; $N'^*(u)$ is the *formal* adjoint of $N'(u)$, and $\Gamma'^*(u)$ includes all boundary operators on v. We denote, symbolically, by δ^*, the distribution associated with the adjoint boundary spaces of the problem (2.1) and (2.2) (see Ref. 3). The exact meaning of δ and δ^* will become clearer in Section 4. The operator $N'^*(u) + \delta^* \cdot \Gamma'^*(u)$ is a matrix whose (ij)-th element is the operator $[\partial(N_j + \delta \cdot \Gamma_j)/\partial u_i]^*$.

Next, we define the operators

$$L(u)h \equiv \int_0^1 N'(\epsilon u)h d\epsilon , \quad \gamma(u)h \equiv \int_0^1 \Gamma'(\epsilon u)h d\epsilon , \tag{2.7}$$

and

$$L^*(u)v \equiv \int_0^1 [N'(\epsilon u)]^* \, v d\epsilon, \quad \gamma^*(u)v \equiv \int_0^1 [\Gamma'(\epsilon u)]^* \, v d\epsilon . \tag{2.8}$$

Note that the operators L, γ, L^*, γ^* retain a nonlinear parametric dependence on u but L and γ act linearly on h while L^* and γ^* act linearly on v. In view of their definitions given in Eq. (2.7), the operators L and γ are related to the original nonlinear operators N and Γ via the important relationship

$$[L(u) + \delta \cdot \gamma(u)] \, u = N(u) + \delta \cdot \Gamma(u) . \tag{2.9}$$

We note that if u happens to be the actual solution of (2.1)-(2.2) then $(L(u)+\delta \cdot \gamma(u))u = f + \delta \cdot g$. This relationship underscores the important fact that, in contradistinction to the variational operators $N'(u)$ and $\gamma'(u)$, it is the pair of *integrated* operators $\{L(u), \gamma(u)\}$ which restores exactly the original nonlinear system (2.1) and (2.2) when applied to u. It also follows from Eqs. (2.6), (2.7) and (2.8) that the (ij)-th component of the (matrix) operator $L^*(u) + \delta^* \cdot \gamma^*(u)$ is obtained by taking the formal adjoint of the (ji)-th component of $L(u) + \delta \cdot \gamma(u)$, i.e.

$$\left(L^*(u) + \delta^* \cdot \gamma^*(u)\right)_{ij} = \left[\left(L(u) + \delta \cdot \gamma(u)\right)_{ji}\right]^* . \tag{2.10}$$

We now introduce the backward (retarded) and forward (advanced) propagators $G(u)$ and $G^*(u)$ as the respective inverses of the operators $L(u) + \delta \cdot \gamma(u)$ and $L^*(u) + \delta^* \cdot \gamma^*(u)$, i.e.,

$$[L(u) + \delta \cdot \gamma(u)] \, G(u) = 1 \tag{2.11}$$

and

$$[L^*(u) + \delta^* \cdot \gamma^*(u)] \, G^*(u) = 1 , \tag{2.12}$$

where 1 denotes the unit operator. The propagators $G^*(u)$ and $G(u)$ are $n \times n$ matrices whose components are operators on $L_2(\Omega)$. In terms of formal integral kernels, Eqs. (2.11) and (2.12) can be written as

$$[L(u(x)) + \delta \cdot \gamma(u(x))]\; G(u(x);\, x,\, x') = \delta(x - x') \tag{2.13}$$

and

$$[L^*(u(x)) + \delta^* \cdot \gamma^*(u(x))]\; G^*(u(x);\, x,\, x') = \delta(x - x') \tag{2.14}$$

where $x \in \bar{\Omega}$, $x' \in \bar{\Omega}$.

The forward and backward propagators are related to each other through the reciprocity relationship

$$G^*(u(x);\, x,\, x') = G(u(x');\, x',\, x). \tag{2.15}$$

This reciprocity relationship can be readily derived by taking the inner products of Eqs. (2.13) and (2.14) with $G^*(u)$ and $G(u)$, respectively, and using Eq. (2.10).

The solution u of the original nonlinear system can now be expressed in terms of either $G^*(u)$ or $G(u)$. In terms of the forward propagator $G^*(u)$, for example, u is obtained as follows:

$$\begin{aligned}
u &= < u, \delta > \\
&= < u, \delta > - < N(u) + \delta \cdot \Gamma(u),\, G^*(u) > + < f + \delta \cdot g,\, G^*(u) > && \text{[by Eq. (2.4)]} \\
&= < u, \delta > - < [L(u) + \delta \cdot \gamma(u)]\, u,\, G^*(u) > + < f + \delta \cdot g,\, G^*(u) > && \text{[by Eq. (2.9)]} \\
&= < u, \delta > - < u,\, [L^*(u) + \delta^* \cdot \gamma^*(u)] G^*(u) > + < f + \delta \cdot g,\, G^*(u) > && \text{[by Eq. (2.10)]} \\
&= < f + \delta \cdot g, G^*(u) > . && \text{[by Eq. (2.14)]}
\end{aligned}$$

In view of the reciprocity relationship (2.15), the solution u can thus be written in terms of either $G^*(u)$ or $G(u)$ as

$$u \;= < f + \delta \cdot g,\, G^*(u) > \;=\; < G(u),\, f + \delta \cdot g >, \tag{2.16}$$

or

$$u(x) = \int_\Omega G^*\,(u(x');\, x',\, x)\,(f(x') + \delta \cdot g(x'))\; dx' \tag{2.17}$$

and

$$u(x) = \int_\Omega G(u(x);\, x,\, x')\,(f(x') + \delta \cdot g(x'))\; dx' \,. \tag{2.18}$$

A formal proof of the uniqueness of the representations (2.17) and (2.18) can be constructed by considering two equivalent forms of Eq. (2.4), namely

$$N_1(u) + \delta \cdot \Gamma_1(u) = f_1 + \delta \cdot g_1 \tag{2.19}$$

and

$$N_2(u) + \delta \cdot \Gamma_2(u) = f_2 + \delta \cdot g_2\,, \tag{2.20}$$

where the system (2.20) is obtained from (2.19) by a transformation that *(i)* vanishes when applied to the unique solution $\tilde{u}$ of Eq. (2.4), and *(ii)* does not introduce new solutions in addition to $\tilde{u}$. Note that $\tilde{u}$ must be a solution branch free of bifurcation points so that the required Gâteaux-derivatives of N_1, N_2, Γ_1, and Γ_2 exist. Thus the counterparts of Eq. (2.5) through (2.9) exist for both Eqs. (2.19) and (2.20); therefore Eqs. (2.19) and (2.20) can be written as

$$[L_1(u) + \delta \cdot \gamma_1(u)]u = f_1 + \delta \cdot g_1 \tag{2.21}$$

and

$$[L_2(u) + \delta \cdot \gamma_2(u)]u = f_2 + \delta \cdot g_2 \,. \tag{2.22}$$

The forward propagators corresponding to Eqs. (2.21) and (2.22) are then defined as the solutions of

$$[L_1(\tilde{u}) + \delta \cdot \gamma_1(\tilde{u})]\, G_1(\tilde{u}) = 1 \tag{2.23}$$

and

$$[L_2(\tilde{u}) + \delta \cdot \gamma_2(\tilde{u})]\, G_2(\tilde{u}) = 1 \,, \tag{2.24}$$

respectively. Thus, Eqs. (2.19) and (2.20) will be equivalent if we show that the solution can be obtained by using alternatively $G_1(\tilde{u})$ or $G_2(\tilde{u})$, i.e.,

$$\tilde{u} = < G_1(\tilde{u}),\, f_1 + \delta \cdot g_1 > \; = \; < G_2(\tilde{u}),\, f_2 + \delta \cdot g_2 > \,. \tag{2.25}$$

The notation for the proof of (2.25) can be simplified considerably by simply interpreting $G_1(\tilde{u})$ and $G_2(\tilde{u})$ as operators such that Eq. (2.25) can be written as

$$\tilde{u} = G_1(\tilde{u})(f_1 + \delta \cdot g_1) = G_2(\tilde{u})(f_2 + \delta \cdot g_2) \,. \tag{2.26}$$

Using the Dyson formula, relating the resolvents of perturbed and unperturbed operators,

$$G_2(\tilde{u}) = G_1(\tilde{u}) - G_1(\tilde{u})\, [L_2(\tilde{u}) + \delta \cdot \gamma_2(\tilde{u}) - L_1(\tilde{u}) - \delta \cdot \gamma_1(\tilde{u})]\, G_2(\tilde{u}) \,,$$

we can formally prove the validity of Eq. (2.26) as follows:

$$\begin{aligned}
\tilde{u} &= G_2(\tilde{u})(f_2 + \delta \cdot g_2) \\
&= G_1(\tilde{u})(f_2 + \delta \cdot g_2) - G_1(\tilde{u})\, [L_2(\tilde{u}) + \delta \cdot \gamma_2(\tilde{u}) - L_1(\tilde{u}) - \delta \cdot \gamma_1(\tilde{u})]\; G_2(\tilde{u})(f_2 + \delta \cdot g_2) \\
&= G_1(\tilde{u})(f_2 + \delta \cdot g_2) - G_1(\tilde{u})\, [L_2(\tilde{u}) + \delta \cdot \gamma_2(\tilde{u}) - L_1(\tilde{u}) - \delta \cdot \gamma_1(\tilde{u})]\, \tilde{u} \\
&= G_1(\tilde{u})(f_2 + \delta \cdot g_2) - G_1(\tilde{u})\, [L_2(\tilde{u}) + \delta \cdot \gamma_2(\tilde{u})]\, \tilde{u} + G_1(\tilde{u})\, [L_1(\tilde{u}) + \delta \cdot \gamma_1(\tilde{u})]\, \tilde{u} \\
&= G_1(\tilde{u})(f_2 + \delta \cdot g_2) - G_1(\tilde{u})(f_2 + \delta \cdot g_2) + G_1(\tilde{u})(f_1 + \delta \cdot g_1) \\
&= G_1(\tilde{u})(f_1 + \delta \cdot g_1) \,,
\end{aligned} \tag{2.27}$$

which proves that the two representations (2.19) and (2.20) are indeed equivalent. We close this Section with the following remark. The operators $L(u)$, $L^*(u)$, $\gamma(u)$, $\gamma^*(u)$ are canonically and univocally determined from the original operators $N(u)$ and $\Gamma(u)$. Starting from a different pair N, Γ we would obtain, in principle, different operators L, L^*, γ, γ^*. This happens even if the two original systems admit the same solution $\tilde{u}$, e.g.

$$N(u) + \delta \cdot \Gamma(u) = f + \delta \cdot g \tag{2.28}$$

$$N(u) + F(u) + \delta \cdot \Gamma(u) + \delta \cdot \Phi(u) = f + F(\tilde{u}) + g + \delta \cdot \Phi(\tilde{u}) \tag{2.29}$$

It can be proved [2] that the two equivalent forms lead to the same solution. The proof is analogous to the proof indicated in (2.27) and will be illustrated in Example 4.1.

3. INTEGRAL EQUATIONS FOR PROPAGATORS

It is possible and useful for practical applications (as will be shown in Section 4) to derive relationships between the forward and backward propagators $G^*(u^o)$ and $G(u^o)$ corresponding to a known vector $u^o \in L_2^n(\Omega)$ and the forward and backward propagators $G^*(u^o)$ and $G(u^o)$ corresponding to the solution (assumed unique) of the original nonlinear system (2.1) and (2.2). These relationships can be readily obtained by forming the respective inner products of $G^*(\tilde{u})$ with

$$[L(u^o) + \delta \cdot \gamma(u^o)]\, G(u^o) = 1 \tag{3.1}$$

and of $G(u^o)$ with

$$[L(\tilde{u}) + \delta^* \cdot \gamma^*(\tilde{u})]\, G^*(\tilde{u}) = 1 \tag{3.2}$$

and subtracting the latter inner product from the former. This procedure gives the following closed-form nonlinear (in general, integro-differential) equation for the forward propagator $G^*(\tilde{u})$:

$$G^*(\tilde{u}) = G(u^o) + < G(u^o),\, [L^*(u^o) + \delta^* \cdot \gamma^*(u^o) - L^*(\tilde{u}) - \delta^* \cdot \gamma^*(\tilde{u})]\, G^*(\tilde{u}) > \tag{3.3}$$

A similar equation is obtained for the backward propagator $G(\tilde{u})$:

$$G(\tilde{u}) = G^*(u^o) + < [L(u^o) + \delta \cdot \gamma(u^o) - L(\tilde{u}) - \delta \cdot \gamma(\tilde{u})]\, G(\tilde{u}),\, G^*(u^o) > . \tag{3.4}$$

Note that Eqs. (3.3) and (3.4) are exact and their nonlinear features reflect exactly the nonlinearities of the original system. In practice, they may be difficult to solve; therefore, if the propagators $G^*(u)$ and $G(u)$ are not needed *per se*, Eqs. (3.3) and (3.4) do not have to be solved explicitly but can instead be advantageously used to obtain the solution, $\tilde{u}$ as follows:

$$\begin{aligned}
\tilde{u} &= < \tilde{u}, \delta > \\
&= < \tilde{u},\, [L^*(u^o) + \delta^* \cdot \gamma^*(u^o)]\, G^*(u^o) > \\
&= < \tilde{u},\, [L^*(u^o) + \delta^* \cdot \gamma^*(u^o) - L^*(\tilde{u}) - \delta^* \cdot \gamma^*(\tilde{u}))]\, G^*(u^o) > \\
&\qquad + < \tilde{u},\, [L^*(\tilde{u}) + \delta^* \cdot \gamma^*(\tilde{u})]\, G^*(u^o) > \\
&= < [L(\tilde{u}) + \delta \cdot \gamma(\tilde{u})]\tilde{u},\, G^*(u^o) > \\
&\qquad + < \tilde{u},\, [L^*(u^o) + \delta^* \cdot \gamma^*(u^o) - L^*(\tilde{u}) - \delta^* \cdot \gamma^*(\tilde{u})]\, G^*(u^o) >
\end{aligned}$$

or, finally,

$$\tilde{u} = < f + \delta \cdot g,\, G^*(u^o) > + < \tilde{u},\, [L^*(u^o) + \delta^* \cdot \gamma^*(u^o) - L^*(\tilde{u}) - \delta^* \cdot \gamma^*(\tilde{u})]\, G^*(u^o) > \tag{3.5}$$

Equation (3.5) is, in general, an integro-differential equation for the solution of the original nonlinear system (2.1) and (2.2) in terms of the known source terms f and g, and the known propagator $G^*(u^o)$. In some important particular cases, such as the Burgers and the Korteweg-de Vries equations, Eq. (3.5) reduces to a purely integral equation that can be solved very efficiently and accurately.[5] This aspect will be discussed further in the next section.

To obtain the solution to the original nonlinear system, the integral equation (3.5) is, in principle, more advantageous to work with because integral equations are easier to analyze and solve. For example, the contraction principle and fixed point theorems could be more readily applied to the integral form (3.5) – to analyze questions of existence and uniqueness – but can seldom be used directly on a nonlinear initial/boundary value problem such as (2.1) and (2.2). Furthermore, computational and numerical analysis methods are comparatively more developed for (and easier to implement on) integral equations than for differential ones.

4. ILLUSTRATIVE EXAMPLES

4.1 The Riccati Equation

A simple illustration that nevertheless highlights the main features of applying the general formalism presented in Sections 2 and 3 is provided by considering the scalar Riccati equation

$$N(u) = \frac{du}{dt} + bu^2 - c = 0, \quad t \in (0, t_f), \qquad b > 0,\ c > 0\,, \tag{4.1}$$

subject to the initial condition

$$u(0) = u_i > 0\,. \tag{4.2}$$

Solving this Riccati equation in $L_2(0, t_f)$ by standard procedures gives the unique solution

$$\tilde{u}(t) = \frac{u_i + \sqrt{c/b}\ \tanh(t\sqrt{bc})}{1 + u_i\sqrt{b/c}\ \tanh(t\sqrt{bc})}\,. \tag{4.3}$$

Comparison of Eqs. (4.1) and (4.2) with Eq. (2.4) leads to the identifications $\bar{\Omega} \to [0, t_f]$, $\Gamma(u) \to u(t)$, $f \to c$, and $g \to u_i$, respectively. The Gâteaux-derivative of $N(u)$ in Eq. (4.1) is $N'(u)h = \frac{dh}{dt} + 2buh$, which leads to the following expressions for the integrated variational operators $L(u)$ and $L^*(u)$:

$$L(u)h = \frac{dh}{dt} + buh \tag{4.4}$$

and

$$L^*(u)v = -\frac{dv}{dt} + buv\,. \tag{4.5}$$

The backward and forward propagator kernels $G(u(t); t, t')$ and $G^*(u(t); t, t')$ are thus defined, respectively, as the solutions of

$$\begin{cases} L(u)G(u(t);\, t,\, t') = \frac{dG_u}{dt} + bu\, G_u = \delta(t - t') \\ G(u(t);\, t,\, t') = 0\,, \qquad \text{for } t < t',\ \text{in particular for } t = 0 \end{cases} \tag{4.6}$$

and

$$\begin{cases} L^*(u)G^*(u(t);\, t,\, t') = -\frac{dG^*_u}{dt} + bu\, G^*_u = \delta(t-t') \\ G^*(u(t);\, t;\, t') = 0\,, \qquad \text{for } t > t'\,, \text{ in particular for } t = t_f. \end{cases} \tag{4.7}$$

The forward propagator is readily computed as:

$$G^*(u(t);\, t,\, t') = H(t'-t)\, \exp\left[\int_{t'}^{t} bu(\tau)\, d\tau\right]\,, \tag{4.8}$$

where

$$H(t'-t) = \begin{cases} 0, & t' < t \\ 1, & t' \geq t \end{cases} \tag{4.9}$$

is the Heaviside function. An analogous expression is obtained for $G(u(t); t, t')$.

The solution $u(t)$ of the Riccati equation is obtained in terms of the forward propagator $G^*(u)$ as

$$u(t) = \int_0^{t_f} cG^*(u(t');\, t',\, t)dt' + u_i G^*(u_i;\, 0, t)\,, \tag{4.10}$$

or, on substituting the respective expression for $G^*(u)$,

$$u(t) = \left[u_i + c\int_0^t dt'\, \exp\left(\int_0^{t'} bu(\tau)\, d\tau\right)\right] \times \exp\left(-\int_0^t bu(\tau)\, d\tau\right)\,. \tag{4.11}$$

The nonlinear integral equation (4.11) represents a fixed-point form of the original Riccati equation. It is easy to verify that the solution $u(t)$ given by Eq. (4.11) indeed satisfies the original equation (4.1) and the initial condition (4.2).

In order to test the advantages of the propagators method, three numerical methods have been applied [4] to solve Eqs. (4.1) and (4.2):

1. an iterative scheme has been applied directly to the original equation (4.1) and (4.2) in the form

$$u_{n+1}(t) = \int_0^t [c - bu_n^2(s)]ds + u_i \qquad\qquad u_o(t) = u_i \tag{4.12}$$

2. an iterative scheme has been applied to the propagator expression (4.11)

$$u_{n+1}(t) = \exp\left[-\int_0^t bu_n(\tau)\right] \times \left\{u_i + c\int_0^t d\tau\, \exp\left(\int_0^\tau bu_n(y)dy\right)\right\} \tag{4.13}$$

3. and, finally, an implicit finite difference method.

The propagator method ranked first in terms of stability and computer efficiency and was close to the finite difference method in terms of accuracy. The greater stability is due to the absence of secular terms in (4.13) as opposed to (4.12).

Recalling the expression of $u(t)$ from Eq. (4.3), we note that Eqs. (4.1) and (4.2) can equivalently be written as

$$\frac{du}{dt} + bu\tilde{u} - c = 0 \tag{4.1'}$$

$$u(0) = u_i, \tag{4.2'}$$

or

$$\frac{du}{dt} + b\tilde{u}^2 - c = 0 \tag{4.1''}$$

$$u(0) = u_i. \tag{4.2''}$$

The propagators corresponding to Eqs. (4.1′) and (4.2′) will, of course, differ from those corresponding to Eqs. (4.1″) and (4.2″), and they will all differ from those corresponding to Eqs. (4.6) and (4.7). For example, the forward propagator $G_1^*(u(t);\, t,\, t')$ corresponding to the system (4.1′) and (4.2′) is the solution of

$$\begin{cases} -\frac{dG_1^*}{dt} + b\tilde{u}\, G_1^* = \delta(t-t') \\ G_1^*(u(t);t,t') = 0 \qquad\qquad t > t' \end{cases} \tag{4.14}$$

while the forward propagator $G_2^*(u(t);\, t,\, t')$ corresponding to the system (4.1″) and (4.2″) is the solution of

$$\begin{cases} -\frac{dG_2^*}{dt} = \delta(t-t') \\ G_2^*(u(t);t,t') = 0 \qquad\qquad t > t' \end{cases} . \tag{4.15}$$

The explicit expressions of G_1^* and G_2^* are

$$G_1^*(u(t);t,t') = H(t'-t)\, \exp\left[\int_{t'}^{t} b\tilde{u}(\tau)\, d\tau\right], \tag{4.16}$$

and

$$G_2^*(u(t);t,t') = H(t'-t)\,, \tag{4.17}$$

respectively. Note that, in contradistinction to $G^*(u(t);\, t,\, t')$ given by (4.8), neither G_1^* nor G_2^* depend on $u(t)$. Although G^*, G_1^*, and G_2^* are clearly distinct from each other, they nevertheless yield the same solution $u(t) = \tilde{u}(t)$ when substituted in the adequate realization of Eq. (2.17), namely

$$\begin{aligned} u(t) &= \int_0^{t_f} G^*(u(t');\, t',\, t)\, (c + \delta(t') \cdot u_i)\, dt' = \int_0^{t_f} G_1^*(u(t');t',t)\, (c + \delta(t') \cdot u_i)\, dt' \\ &= \int_0^{t_f} G_2^*(u(t');t',t)\, \left[-b\tilde{u}^2(t') + c + \delta(t') \cdot u_i\right]\, dt' = \tilde{u}(t). \end{aligned} \tag{4.18}$$

The equivalences in (4.18) illustrate the fact that all pairs of operators $\{L_i^*,\, \gamma_i^*\}$ are equivalent to the pair $\{L^*,\, \gamma^*\}$ obtained canonically as described in Section 2 in that the respective propagators G_i^*, although distinct from each other, all lead to the same solution $\tilde{u}(t)$ of the original nonlinear system. A similar statement is valid, of course, for the pairs $\{L_i, \gamma_i\}$ and the corresponding backward propagators G_i.

4.2 The Homogeneous Carleman Model

In the kinetic theory of gases, the homogeneous Carleman model gives the time-dependent concentrations $u_1(t)$ and $u_2(t)$ of two gases as the solutions of the following nonlinear system:

$$\begin{cases} \frac{du_1}{dt} + u_1^2 - u_2^2 + \delta(t) \cdot u_1 = \delta(t) \cdot u_{1,0} \\ \frac{du_2}{dt} + u_2^2 - u_1^2 + \delta(t) \cdot u_2 = \delta(t) \cdot u_{2,0} \end{cases} \tag{4.19}$$

where $u_{1,0}$ and $u_{2,0}$ are the respective initial concentrations, and $\delta(t)$ is, as before, the δ-distribution.

In the real space $L_2(0, t_f) \times L_2(0, t_f)$, the solution of this Carleman system is readily obtained as

$$u_1(t) = \frac{1}{2}\left[c + (u_{1,0} - u_{2,0})e^{-2ct}\right], \tag{4.20}$$

and

$$u_2(t) = \frac{1}{2}\left[c - (u_{1,0} - u_{2,0})e^{-2ct}\right], \tag{4.21}$$

where the constant c is defined as

$$c = u_{1,0} + u_{2,0}. \tag{4.22}$$

For illustrative purposes, we will solve the Carleman system using the backward rather than the forward propagator. Since

$$N'(u) + \delta \cdot \Gamma'(u) = \begin{pmatrix} \frac{d}{dt} + 2u_1 & -2u_2 \\ -2u_1 & \frac{d}{dt} + 2u_2 \end{pmatrix} + \delta(t) \cdot \begin{pmatrix} 1 & 0 \\ 0 & 1 \end{pmatrix}, \tag{4.23}$$

it follows that the backward propagator

$$G(u(t); t, t') = \begin{pmatrix} G_{11} & G_{12} \\ G_{21} & G_{22} \end{pmatrix} \tag{4.24}$$

is the solution of

$$\begin{cases} \begin{pmatrix} \frac{d}{dt} + u_1 & -u_2 \\ -u_1 & \frac{d}{dt} + u_2 \end{pmatrix} \begin{pmatrix} G_{11} & G_{12} \\ G_{21} & G_{22} \end{pmatrix} = \delta(t - t_1) \cdot \begin{pmatrix} 1 & 0 \\ 0 & 1 \end{pmatrix} \\ G_{ij}(u(t); t, t') = 0; \quad t < t'; \quad i, j = 1, 2. \end{cases} \tag{4.25}$$

Solving Eq. (4.25) gives

$$G_{11}(u(t); t, t') = H(t - t')\left[e^{c(t'-t)} + \int_{t'}^{t} e^{c(\tau - t)} u_2(\tau) d\tau\right] \tag{4.26}$$

$$G_{21}(u(t); t, t') = H(t - t')\left[1 - e^{c(t'-t)} \int_{t'}^{t} e^{c(\tau - t)} u_2(\tau) d\tau\right] \tag{4.27}$$

$$G_{12}(u(t);t,t') = H(t-t')\left[1 - e^{c(t'-t)}\int_{t'}^{t} e^{c(\tau-t)}u_1(\tau)d\tau\right] \tag{4.28}$$

$$G_{22}(u(t);t,t') = H(t-t')\left[e^{c(t'-t)} + \int_{t'}^{t} e^{c(\tau-t)}u_1(\tau)d\tau\right] \tag{4.29}$$

Using Eqs. (4.26) through (4.29) in Eq. (2.18) gives the solution $u = (u_1, u_2)$ of the Carleman system as:

$$u_1(t) = u_{1,0}G_{11}(u(t);t,0) + u_{2,0}G_{12}(u(t);t,0) = \frac{1}{2}\left[c + (u_{1,0} - u_{2,0})e^{-2ct}\right] \tag{4.30}$$

and

$$u_2(t) = u_{1,0}G_{21}(u(t);t,0) + u_{2,0}G_{22}(u(t);t,0) = \frac{1}{2}\left[c - (u_{1,0} - u_{2,0})e^{-2ct}\right] \tag{4.31}$$

As expected, Eqs. (4.30) and (4.31) are identical to Eqs. (4.20) and (4.21), respectively. Finally, we note that, just like in the illustrations presented in Section 4.1, one could construct alternative representations of the Carleman system, leading to alternative expressions for the respective propagators. Ultimately, of course, all of these alternative representations lead to the same solution u, as given by (4.20) and (4.21); this is illustrated in more detail in Ref. 2.

4.3 A Generalized Burgers-Korteweg-deVries Equation

The Burgers and Korteweg-deVries equations can be viewed as particular cases of the more general equation:

$$u_t + \eta u_x + u^p u_x - \nu u_{xx} + \epsilon u_{xxx} = 0\,, \tag{4.32}$$

where $u(x,t)$ is a real-valued function of the spatial variable x and time variable t. In Eq. (4.32), the subscripts denote partial differentiations, and the quantities η, p, ν, ϵ are positive real constants. We consider $x \in (0,1)$, $t \in (0,t_f)$, $u \in L^2((0,1)\times(0,t_f))$ and the inner product $< u,v >= \int_0^1 dx \int_0^{t_f} u(x,t)v(x,t)dt$. We seek the solution of Eq. (4.32) subject to the periodic boundary conditions

$$\left\{\partial^j u/\partial x^j\right\}_{x=0} = \left\{\partial^j u/\partial x^j\right\}_{x=1} \qquad j = 0,1,2, \tag{4.33}$$

and initial condition

$$u(x,0) = u_i(x)\,. \tag{4.34}$$

The initial-boundary value problem (4.32) - (4.34) has been solved[5] for arbitrary values of the positive constants η, p, ν, and ϵ, yet in the following discussion we shall restrict ourselves to the Korteweg-deVries equation by taking $\eta = 0$, $\nu = 0$, and $p = 1$. In this case, the integral equation for $u(x,t)$, *viz.* Eq. (3.5), becomes[5]

$$u(x,t) = u_i(x')G_o^*(u(x',0);x',0;\,x,t)\,dx'$$

$$-\int_0^t\int_0^1 dt'dx'\left\{u(x',t')u^o - u^2(x',t')\right\}\,\partial G_o^*(u(x',t');x',t';\,x,t)/\partial x' \tag{4.35}$$

where u^o is a known (user-selected, *cf.* Section 3) constant, and the auxiliary propagator $G^*(u^o; x, t, x', t')$ corresponding to u^o is the solution of

$$\begin{cases} L^*(u^o)G^*(u^o; x,t,x',t') \equiv \left(-\dfrac{\partial}{\partial t} - \dfrac{u^o}{2}\dfrac{\partial}{\partial x} - \epsilon\dfrac{\partial^3}{\partial x^3}\right) G^*(u^o; x,t,x',t') \\ \qquad = \delta(x-x')\delta(t-t') \\ G^*(u^o; x,t,x',t') = 0 \qquad \text{for } t > t' \\ \{\partial^j G^*/\partial x^j\}_{x=0} = \{\partial^j G^*/\partial x^j\}_{x=1} \end{cases} \tag{4.36}$$

The linear system (4.36) can readily be solved to obtain the explicit expression

$$G^*(u^o; x,t,x',t') = H(t-t')\left\{1 + 2\sum_{n=1}^{\infty} \cos 2\pi n\left[(u^o/2 - 4\pi^2 n^2\epsilon)(t-t') - (x-x')\right]\right\}. \tag{4.37}$$

A benchmark traditionally used for testing numerical methods is the solitary wave solution

$$u(x,t) = A\,\text{sech}^2\left[\sqrt{A/12\epsilon}\,(x - 1/2 - t/3)\right] \tag{4.38}$$

of the Korteweg-deVries equation; this solution corresponds to the initial condition

$$u_i(x) = A\,\text{sech}^2\left[\sqrt{A/12\epsilon}\,(x-0.5)\right] \tag{4.39}$$

where A is the soliton amplitude. For this benchmark, the method presented in Sections 2 and 3, termed the ***propagator method*** in the table below, has been compared to three leading methods for solving nonlinear equations: a finite-difference method as reported by Zabuski and Kruskal,[6] an inverse scattering transform (IST) method as reported by Taha and Ablowitz,[7] and a Galerkin finite element method as reported by Dougalis and Karakashian.[8] Results for $A = 2$, $\epsilon = 1.042 \times 10^{-4}$, and $t_f = 0.15$ are tabulated below. For the propagator method in the table below, the number of spatial steps is replaced by the number of Fourier modes used.

Method	**ℓ_∞-error$\times 10^3$**	**No. of time steps**	**No. of spatial steps**	**CPU(sec)**
Finite-difference	3.31	25,000	840	22.63
IST	3.32	10	400	0.300
Finite-element	3.105	20	128	0.284
Propagator	2.95	90	36	0.262

These results indicate that the propagator method is at least as efficient and accurate as the special purpose IST method and the finite-element method, while being considerably simpler to implement.

5. CONCLUSIONS

We have presented in this paper a canonical formalism for solving nonlinear initial/boundary value problems in terms of propagators that generalize the customary Green's functions of linear theory. This formalism is not limited to special boundary and/or initial conditions, or to special structures of the underlying nonlinear operators. Fundamental to the development of our formalism are the operators $L(u)$ and $L^*(u)$ obtained by a functional integration of the Gâteaux-derivative $N'(u)$ and of its adjoint $[N'(u)]^*$, respectively. It is important to note that, in contradistinction to $N'(u)$, it is the operator $L(u)$ and its corresponding (integrated variational) boundary operator $\gamma(u)$ that satisfy the relationship $[L(u) + \delta \cdot \gamma(u)]u = N(u) + \delta \cdot \Gamma(u)$. Consequently, the forward and backward propagators $G^*(u)$ and $G(u)$, which are solutions of the equations involving $L^*(u)$ and $L(u)$, respectively, contain all the information needed to solve the original nonlinear system. This is the very reason that the propagators $G^*(u)$ and $G(u)$ generalize the customary Green's functions from linear theory to nonlinear systems. It is readily apparent from the development of our formalism that when the original system is linear, the operators L^* and L become independent of u, so $G^*(u)$ and $G(u)$ become the customary (u-independent) Green's functions.

We have shown that the forward and backward propagators $G^*(u)$ and $G(u)$ satisfy a reciprocity relationship analagous to that satisfied by the customary Green's functions. We have further shown that these propagators contain nonlinearities stemming from the integro-differential equations they satisfy. In contradistinction to nonlinearities that may appear, for example, in the Green's functions in many-body and field theories (where such nonlinearities are not intrinsic to the respective theories but are introduced as a result of closure approximations), the nonlinearities in the propagators $G^*(u)$ and $G(u)$ reflect *exactly* the nonlinearities present in the original initial/boundary value problem.

The derivations underlying this canonical formalism are formal in the sense that we have not addressed issues such as existence of solutions, well-posedness, or regularity. Such issues cannot be addressed in a framework as general as our formalism, but have to be analyzed in a problem-specific context. Nevertheless, the conversion of a nonlinear boundary/initial value problem into an integral form, which is the *raison d'être* for our formalism, is generally advantageous because it opens the way to *(a)* applying powerful methods generally unavailable for partial differential operators (e.g., fixed-point, contraction principle), and *(b)* easy implementation of accurate and efficient numerical techniques. These advantages have already been highlighted by the promising numerical results for the nonlinear test problems analyzed thus far.

ACKNOWLEDGMENTS

This paper was partially supported by the U. S. Department of Energy under contract DE-AC05-84OR21400 with Martin Marietta Energy Systems, Inc. The participation of V. Protopopescu at the Workshop was made possible by the Italian National Council for Research.

REFERENCES

[1] Cacuci, D. G., Perez, R. B., Protopopescu, V., "Duals and Propagators: A Canonical Formalism for Nonlinear Equations," J. Math. Phys., 29, 353-361 (1988); see also D. G. Cacuci, R. B. Perez, V. Protopopescu, "Extending Green's Function Formalism to General Nonlinear Equations," in *Proceedings of the IVth Workshop on Nonlinear Evolution Equations and Dynamical Systems*, ed. J.J.P. Léon, World Scientific, Singapore, 1988, pp. 87-95.

[2] Cacuci, D. G., Protopopescu, V., "Propagators for Nonlinear Systems," J. Phys. A., (1989), 22 (1989).

[3] Coddington, E. A., Levinson, N., *Theory of Ordinary Differential Equations*, McGraw Hill, New York, 1955.

[4] Azmy, Y. Y., Cacuci, D. G., Protopopescu, V., "A Comparison Between the Methods of Decomposition and Propagators for Nonlinear Equations," unpublished report.

[5] Cacuci, D. G., Karakashian, O., "A New Method for Solving the Generalized Burgers-Korteweg-de Vries Equation," J. Comp. Phys., submitted.

[6] Zabuski, N. J., Kruskal, M. D., "Interactions of "Solitons" in a Collisionless Plasma and the Recurrence of Initial States," Phys. Rev. Lett., 15, 240-243 (1965).

[7] Taha, T. R., Ablowitz, M. J., "Analytical and Numerical Aspects of Certain Nonlinear Evolution Equations. III Numerical Korteweg-de Vries Equation," J. Comp. Phys., 55, 231-253 (1984).

[8] Dougalis, V. A., Karakashian, O., "On Some High-Order Accurate Fully Discrete Galerkin Methods for the Korteweg-de Vries Equation," Math. Comp., 45, 329-345 (1985).

Singularity Formation for Vortex Sheets and Hyperbolic Equations

Russel E. Caflisch
Mathematics Department, UCLA
Los Angeles, CA 90024-1555 USA

ABSTRACT

Singularities are expected to occur in a variety of inviscid incompressible flows, the simplest being on a vortex sheet just preceding roll-up of the sheet. We present a new approach to the vortex sheet problem, in which the Birkhoff-Rott equation is approximated by a system of first order non-linear pde's. The system is solved in an analytic function setting, and singularities occur as branch points for the solution. In this paper, the general method is applied to Burger's equation and to the short time existence problem for a 2x2 system with initial singularities.

1. Introduction

Singularities are known or expected to occur in a variety of inviscid, incompressible flows: Kelvin-Helmholtz instability of a vortex sheet, Rayleigh-Taylor instability of stratified flow, unsteady Prandtl boundary layers (only partly inviscid) and possibly the 3D Euler equations. Such singularities are of great important to mathematical theory and computational methods. They may also be of strong physical interest; for example, singularity formation on a vortex sheet immediately precedes the roll-up of the sheet.

This paper presents a general approach to the study of singularity formation, applied to the Kelvin-Helmholtz instability problem. This approach derives from an earlier approximate method of Derek Moore [11,12]. The vortex sheet equation ((2.1) below) is approximated by a nonlinear system of hyperbolic equations. This system is then analyzed in an analytic function setting.

A general theory is partly developed for nonlinear hyperbolic equations of a complex variable.

Research supported in part by the Air Force Office of Scientific Research under URI grant AFOSR 90-0003, the National Science Foundation through grant NSF-DMS-9005881 and the Alfred P. Sloan Foundation.

The resulting solutions are necessarily multi-valued, in analogy to the well-known multi-valued solutions of Burger's equation. The "turning points" of the multi-valued solutions are branch points on a Riemann surface, which evolves along with the solution as the branch points move and interact. Singularities on the vortex sheet correspond to these branch points.

There are two main results in this paper: A new representation of multi-valued solutions of Burger's equations is developed using analytic continuation. Second, a short time existence result is proved for a 2x2 nonlinear hyperbolic system square-root type singularities in its initial data.

2. Vortex Sheets

A vortex sheet is a curve in 2D inviscid, incompressible flow along which the tangential velocity $\tau \cdot \mathbf{u}$ is discontinuous. Parameterize the sheet by a real variable γ, such that $\partial\gamma/\partial s = \tau \cdot [\mathbf{u}] =$ vortex density, and describe points on the sheet by a complex variable $z(\gamma,t) = x + iy$. The sheet evolves according to the Birkhoff-Rott equation.

$$\partial_t z^*(\gamma,t) = (2\pi i)^{-1} PV \int_{-\infty}^{\infty} (z(\gamma,t) - z(\gamma+\zeta,t))^{-1} d\zeta \tag{2.1}$$

in which $z^*(\gamma) = \overline{z(\overline{\gamma})}$ and the integral is a Cauchy principal value integral. Note that this equation is formally analytic in γ, and that z^* is the analytic extension of $\overline{z}$ from the real γ line. Only real values of γ have physical meaning, but the extension of z to complex γ values will be useful mathematically.

A stationary solution of (2.1) is $z \equiv \gamma$, corresponding to a flat vortex sheet of uniform strength. Consider perturbations of this in the form $z(\gamma,t) = \gamma + s(\gamma,t)$. As is well-known this stationary solution is unstable, and the linearized modes for s have the form

$$\hat{s}_{\pm}(k,t) = (1 \pm i) e^{ik\gamma \pm kt/2} \tag{2.2a}$$

$$= (1 \pm i) e^{ik(\gamma \pm (-i)t/2)} \tag{2.2b}$$

Formula (2.2b) shows that the linearized vortex sheet equation is hyperbolic in the complex variable γ, with characteristic speeds $\pm i/2$. Thus we can expect singularities in s to move approximately at speed $\pm i/2$ in the complex γ- plane as drawn in figure 1. This speed will be only slightly modified if the nonlinearities are not large.

As a first result of this observation, the vortex sheet equations are found to be well-posed for initial data that is analytic in a strip $\{\gamma : |\mathrm{Im}\gamma| < \rho\}$. Existence for a time approximately 2ρ (corresponding to speeds $\pm i/2$) was proved by Caflisch and Orellana [1] following earlier work of Sulem, Sulem, Bardos and Frisch [14]. The point vortex and vortex blob numerical methods were shown to be convergent for a short time under the same conditions [3].

A second, converse result is that singularities may form on the vortex sheet. Derek Moore [11,12] first analyzed singularity formation by an asymptotic expansion, using a system of approximate equations that will be discussed in the next section. He found singularities of the approximate form $z(\gamma) \simeq (\gamma - \gamma_0)^{3/2}$ so that absolute value of the curvature $|\kappa| \simeq |z_{\gamma\gamma}|$ of the sheet becomes infinite at a point. His predictions have been numerically verified by Meiron, Baker and Orszag [10], Krasny [8] and Shelley [13]. In addition Krasny [9] has shown that the vortex sheet rolls-up immediately after the singularity appears.

Rigorous construction of singular solutions for the Birkhoff-Rott equation (2.1) has been performed by Duchon and Robert [6] and Caflisch and Orellana [2]. These solutions can have the form $z(\gamma) \simeq (\gamma - \gamma_0)^{\nu}$ for any $\nu > 1$. The vortex sheet position and strength are plotted in figure 2 for $\nu = 3/2$. Construction of such solutions show the vortex sheet problem to be ill-posed in the Sobolev space H^n for $n > 3/2$ [2,7]. Moore's theory [11,12] predicts that the generic form of the singularity is $\nu = 3/2$. The theory described below seems to verify this but has not yet been completed.

3. Nonlinear Approximation for Vortex Sheets

Here we present an approximation to the Birkhoff-Rott equation (2.1), which is valid even if the sheet has singularities. Denote $f = \partial_\gamma z$ and $g = H[\partial_\gamma z]$ in which H is the Hilbert transform defined by

$$Hf(\gamma) = (\pi i)^{-1} PV \int_{-\infty}^{\infty} f(\gamma + \zeta)\zeta^{-1} d\zeta. \tag{3.1}$$

Equation (2.1) is equivalent to the following system

$$\partial_t f^* = (1/2) f^{-2} \partial_\gamma g + E_1 \tag{3.2a}$$

$$\partial_t g^* = -(1/2) f^{-2} \partial_\gamma f + E_2. \tag{3.2b}$$

Denote $W^{s,p}$ as the Sobolev norm with derivatives up to order s in L^p. The error terms E_i are small in the following sense:

<u>Proposition</u> <u>1.</u> If $f, g \in W^{1/2,p}$ then $E_i \in W^{0,r}$ with $r^{-1} = 2p^{-1}$. Therefore E_i is smoother than $\partial_\gamma f$ and $\partial_\gamma g$.

We derive equation (3.2a) and sketch the proof of this proposition. Differentiate (2.1) and use $f = \partial_\gamma z$ to obtain

$$\partial_t f^*(\gamma) = (2\pi i)^{-1} PV \int_{-\infty}^{\infty} \frac{f(\gamma+\zeta) - f(\gamma)}{(z(\gamma+\zeta) - z(\gamma))^2} d\zeta$$

$$= (2\pi i)^{-1} PV \int_{-\infty}^{\infty} \frac{f(\gamma+\zeta) - f(\gamma)}{\zeta^2} (\zeta' \int_0^\zeta f(\gamma+\zeta')d\zeta')^{-2} d\zeta. \quad (3.3)$$

The dominant singularity in the integral of (3.3) occurs in the first term at $\zeta = 0$. Thus it is a good approximation to evaluate the second term at $\zeta = 0$, at which it is just $f(\gamma)^{-2}$. Also use the formula

$$\partial_\gamma Hf = (\pi i)^{-1} PV \int_{-\infty}^{\infty} (f(\gamma+\zeta) - f(\gamma))\zeta^{-2} d\zeta. \quad (3.4)$$

Then (3.3) becomes

$$\partial_t f^*(\gamma) = f(\gamma)^{-2}(2\pi i)^{-1} PV \int_{-\infty}^{\infty} (f(\gamma+\zeta) - f(\gamma))\zeta^{-2} d\zeta + E_1$$

$$= (1/2) f(\gamma)^{-2} \partial_\gamma Hf + E_1 \quad (3.5)$$

which is just (3.3a) since $Hf = g$. In (3.5) E_1 is the error due to evaluating the second term of the integrand in (3.3) at $\zeta = 0$. This commutator-like error can be shown to be small as in the proposition. Equation (3.3b) is derived similarly.

Now discard the error terms E_i and apply* to (3.3) to obtain equations for f and g as well as f^* and g^*. The resulting system is

$$\partial_t f^* = (1/2) f^{-2} \partial_\gamma g \quad (3.6a)$$

$$\partial_t g^* = -(1/2) f^{-2} \partial_\gamma f \quad (3.6b)$$

$$\partial_t f = (1/2) f^{*-2} \partial_\gamma g^* \tag{3.6c}$$

$$\partial_t g = -(1/2) f^{*-2} \partial_\gamma f^* \tag{3.6d}$$

For this system the condition $f^*(\gamma) = \overline{f(\overline{\gamma})}$ is preserved for all time, if it is true initially. Thus this non-local condition need not be included as part of the evolution. Also note that (3.6b) is not exactly the Hilbert transform of (3.6a) because the error terms E_i have been dropped.

The system (3.6) is a nonlinear hyperbolic system with two double characteristics having speeds $i(2ff^*)^{-1}$. As will be shown in the next two sections, singularities for such a system occur as envelopes for the characteristics.

This description of singularities was first presented in the analysis of Moore [12] and generalized in [4]. He derived an approximate system that is similar but simpler than (3.6), and he used this approximate system to derive the singularity results that discussed in section 2. We believe that Moore's analysis is qualitatively correct, but near the time of singularity formation, the errors terms in his approximation seem to be quantitatively significant. The motivation for the new approximation (3.6) is to include all of significant terms up to an including the time of singularity formation, so that a complete analysis of the singularity can be performed.

Note that only consistency of the approximation (3.6) has been shown; that is the errors in (3.6) have been shown to be small. The resulting evolution of (3.6) has not been shown to be close to that of (1.1).

4. Singularities for Burgers Equation

In describing the formation and propagation of singularities for hyperbolic equations, it is natural to start with Burger's equation $\partial_t f + f\partial_x f = 0$. For initial conditions f_0 as in figure 3a, the solution at some later time is multi-valued as in figure 3b. If the solution f is extended as an analytic function of a complex variable $z = x + iy$, solving

$$\partial_t f + f\partial_z f = 0. \tag{4.1}$$

then the turning points are seen to be branch points and the solution is triple-valued everywhere as in figure 3c.

Additional insight into the solution of the complex Burger's equation (4.1) is gained by looking at the characteristics, which are sketched in figure 4. In region I the three values of f came from three real characteristics. In region II they came from one real characteristic and two complex characteristics.

The branch points are envelopes of the characteristics. For initial data $f(z,t=0)=f_0(z)$, the characteristic of (4.1) starting at z_0 is given by

$$z = z_0 + tf_0(z_0) \tag{4.2}$$

Envelopes of the characteristics, indexed by the parameter z_0, are found by solving $\partial z/\partial z_0 = 0$, i.e.

$$1 + tf'_0(z_0) = 0 \tag{4.3}$$

This shows that the envelope originates either at a singularity of f_0 or at $z_0 = \infty$.

Although the multi-valued solutions of Burger's equations are well known, this representation of them as analytic functions on a Riemann surface is new. An important consequence of this representation is that it allows us to discuss the generic form of singularities for Burger's equation. At such an envelope, the generic form of the solution f is $f \simeq (z - z_0)^{1/2}$. This corresponds precisely to the 3/2 power predicted by Moore, since f is the vortex sheet slope (here z is a complex independent variable unrelated to the vortex sheet).

The analytic representation also suggests a numerical method for computing the singular solutions by "going around" the singularities. Numerical solution of Burger's equation (4.1) was performed by a method that does not use the special structure of the equation. The numerically determined location of the branch point is presented in figure 5, for a succession of values of the discretization parameters showing convergence to the exact solution.

5. Singularities for Nonlinear Hyperbolic System

Consider the following 2×2 system in Riemann invariant form

$$f_t + \lambda(f,g)f_z = 0 \tag{5.1a}$$

$$g_t + \mu(f,g)g_z = 0 \tag{5.1b}$$

The main difference between (5.1) and (4.1) is that there are now two-speeds. Thus we expect that an initial singularity will split into two singularities, one traveling at speed λ and the second at speed μ. If the singularities are quadratic, (i.e. $f \sim \sqrt{z - z_0}$) then a function with two such singularities is expected to be defined on a four sheeted Riemann surface, as for example the function

$$f(z) = \sqrt{z - z_1} + \sqrt{z - z_2}. \tag{5.2}$$

The next theorem describes the solution to (5.1) with an initial quadratic singularity. It shows that the solution is defined on the smallest possible Riemann surface. The initial quadratic singularity splits into two quadratic singularities on each sheet, as drawn in figure 6. Note that they do not necessarily line up over each other as in the simple example (5.2). After splitting initially there is no further splitting of the singularity.

<u>Theorem</u> <u>2.</u> **Consider the equation (5.1) with initial data $f(z,0) = f_0(z)$ and $g(z,0) = g_0(z)$. Suppose that λ and μ are analytic and that $\lambda(f_0, g_0) \neq \mu(f_0, g_0)$ in a neighborhood of $z = 0$. Suppose that f_0 and g_0 have a quadratic (square-root type) singularity at $z = 0$ and are defined on a two-sheeted Riemann surface. Then for a short time and in a neighborhood of $z = 0$, there is a unique solution f, g of (5.1) with this initial data, defined on a four sheeted Riemann surface.**

Note 1. At present the uniqueness is proved only among solutions defined on the same four sheet surface.

2. On each sheet of the surface, there are two branch points $z_1(t)$ and $z_2(t)$ with

$$\dot{z}_1(t) = \lambda(z_1(t), t) \tag{5.3a}$$

$$\dot{z}_2(t) = \mu(z_2(t), t) \tag{5.3b}$$

$$f \simeq \begin{cases} f_1 + c_1(z - z_1(t))^{1/2} & z \simeq z_1 \\ f_2 + c_2(z - z_2(t))^{3/2} & z \simeq z_2 \end{cases} \tag{5.4a}$$

$$g \simeq \begin{cases} g_1 + c_3(z - z_1(t))^{3/2} & z \simeq z_1 \\ g_2 + c_4(z - z_2(t))^{1/2} & z \simeq z_2. \end{cases} \tag{5.4b}$$

3. This result shows that the results for Burger's equation can be extended to a a 2x2 system. The corresponding solution is still analytic on a Riemann surface.

This result is proved by introducing new variables u and v in terms of which f and g are smooth. A natural choice for u and v is $u = \sqrt{z - z_1}$ and $v = \sqrt{z - z_2}$. However, this choice has a singularity when $z_1 = z_2$ at $t = 0$. Unfold this singularity using the time variable t by defining $u = \sqrt{z - z_1 + t}$ and $v = \sqrt{z - z_2 - t}$, i.e.

$$t = u^2 - v^2 + (z_1 - z_2) \tag{5.5a}$$

$$z = u^2 + v^2 + (z_1 + z_2). \tag{5.5b}$$

It turns out to work better if this choice is modified to be

$$t = u^2\alpha(u) - v^2\alpha(v) \tag{5.6a}$$

$$z = u^2(1 + \beta(u)) + v^2(1 - \beta(v)) \tag{5.6b}$$

with functions α and β that will be determined as part of the problem.

The system (5.1) is then rewritten as

$$\begin{aligned} &2v\,\{(1 - \beta - v\beta'/2) + \lambda(\alpha + v\alpha'/2)\}f_u \\ &\quad - 2u\,\{(1 + \beta + u\beta'/2) - \lambda(\alpha + u\alpha'/2)\}f_v = 0 \end{aligned} \tag{5.7a}$$

$$\begin{aligned} &2v\,\{(1 - \beta - v\beta'/2) + \mu(\alpha + v\alpha'/2)\}g_u \\ &\quad - 2u\,\{(1 + \beta + u\beta'/2) - \mu(\alpha + u\alpha'/2)\}g_v = 0. \end{aligned} \tag{5.7b}$$

Think of the u variable as approximately parameterizing the λ characteristics and the v variable the μ characteristics. Then (5.7) should be solved for f_u and g_v as

$$f_u = \frac{(1 + \beta + u\beta'/2) - \lambda(\alpha + u\alpha'/2)}{(1 - \beta - v\beta'/2) + \lambda(\alpha + v\alpha'/2)}\,\frac{u}{v}f_v \tag{5.8a}$$

$$g_v = \frac{(1 - \beta - v\beta'/2) + \mu(\alpha + v\alpha'/2)}{(1 + \beta + u\beta'/2) - \mu(\alpha + u\alpha'/2)}\,\frac{v}{u}g_u. \tag{5.8b}$$

Singularities in the system (5.8) are avoided by requiring that

$$(1 + \beta + u\beta'/2) - \lambda(\alpha + u\alpha'/2) = 0 \quad \text{for } v = 0 \tag{5.9a}$$

$$(1 - \beta - v\beta'/2) + \mu(\alpha + v\alpha'/2) = 0 \quad \text{for } u = 0 \tag{5.9b}$$

which implies that the term $v = 0$ in the denominator of (5.8a) is canceled by the first term in the numerator. The other term in the denominator of (5.8a) does not vanish. It follows from (5.9) that

$$\beta(u) = 2u^{-2}\int_0^u \frac{\lambda_0 + \mu_0}{\lambda_0 - \mu_0} u' du' \tag{5.10a}$$

$$\alpha(u) = 2u^{-2}\int_0^u (\lambda_0 - \mu_0)^{-1} u' du' \tag{5.10b}$$

in which $\lambda_0(u') = \lambda(u = u', v = 0), \mu_0(u') = \mu(u = 0, v = u')$. Equations (5.8) and (5.10) are the final system that is to be solved.

Initial data for f and g is specified on $u = v$, corresponding to $t = 0$ by (5.6a), on which $z = 2u^2$. Since f_0 and g_0 have a square root singularity initially, they are smooth functions of the variable u.

With this initial data the system (5.8) and (5.10) has a unique solution that is analytic in u and v, in a small neighborhood of $u = v = 0$. Construction of the solution uses the Cauchy-Kowalewski Theorem. This solution is mapped back to the independent variables z and t by inverting the relationship (5.6). The singularities of this mapping, and hence of f and g, occur at $u = 0$ and at $v = 0$. Furthermore the mapping is multi-valued, since for each z and t there are (locally) four values of u and v. Thus f and g are defined on a four-sheeted Riemann surface in z for any fixed value of t. This concludes the proof of the Theorem.

Much further work is needed to complete this theory. The most important additional phenomenon is the collision between singularities, which is currently being analyzed.

REFERENCES

1. Caflisch, R.E. and Orellana, O.F. "Long time existence for a slightly perturbed vortex sheet," CPAM 39 (1986) 807-838.

2. Caflisch, R.E. and Orellana, O.F. "Singularity formulation and ill-posedness for vortex sheets," SIAM J. Math. Anal. 20 (1989) 293-307.

3. Caflisch, R.E. and Lowengrub, J. "Convergence of the Vortex Method for Vortex Sheets," SIAM J. Num. Anal. 26 (1989) 1060-1080.

4. Caflisch, R.E., Orellana, O.F. and Siegel, M. "A localized approximation for vortical flows," SIAM J. Appl. Math. to appear.

5. Caflisch, R.E., Hou, T. and Ercolani, N., in preparation.

6. Duchon, J., and Robert R., "Global Vortex Sheet Solutions of Euler Equations in the Plan," Comm. PDE to appear.

7. Ebin, D., "Ill-Posedness of the Rayleigh-Taylor and Helmholtz Problems for Incompressible Fluids," Comm. PDE 13 1265-1295.

8. Krasny, R. "On Singularity Formation in a Vortex Sheet and the Point Vortex Approximation," J. Fluid Mech. 167 (1986) 65-93.

9. Krasny, R. "Desingularization of Periodic Vortex Sheet Roll-Up," J. Comp. Phys. 65 (1986) 292-313.

10. Meiron, D.I., Baker, G.R. and Orszag, S.A., "Analytic Structure of Vortex Sheet Dynamics, Part 1, Kelvin-Helmholtz Instability," J. Fluid Mech. 114 (1982) 282-298.

11. Moore, D.W., "The Spontaneous Appearance of a Singularity in the Shape of an Evolving Vortex Sheet," Proc. Roy. Soc. London A 365 (1979) 105-119.

12. Moore, D.W., "Numerical and Analytical Aspects of Helmholtz Instability," in Theoretical and Applied Mechanics, Proc. XVI ICTAM, eds. Niordson and Olhoff, North-Holland, 1984, pp. 629-633.

13. Shelley, M. "A Study of Singularity Formation in Vortex Sheet Motion by a Spectrally Accurate Vortex Method" J. Fluid Mech. to appear.

14. Sulem, C., Sulem, P.L., Bardos, C., and Frisch, U., "Finite Time Analyticity for the Two and Three Dimensional Kelvin-Helmholtz Instability," Comm. Math. Phys. 80 (1981) 485-516.

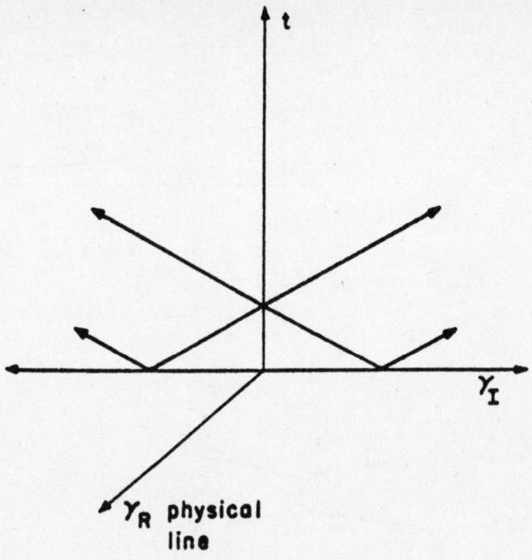

Figure 1. Linear characteristics with speeds $\pm i/2$.

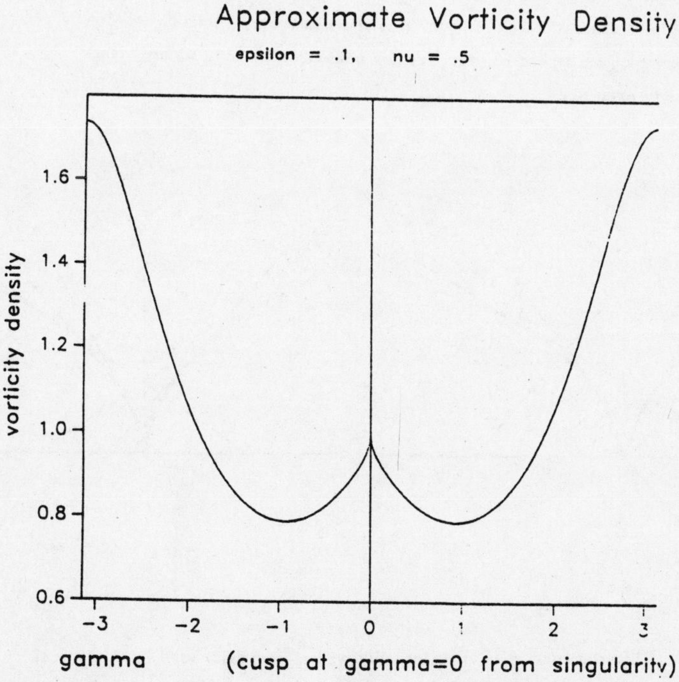

Figure 2. Vortex sheet strength $\sigma = |\partial z/\partial\gamma|^{-1}$, showing a singularity in $\partial^2 z/\partial\gamma^2$ at $\gamma = 0$.

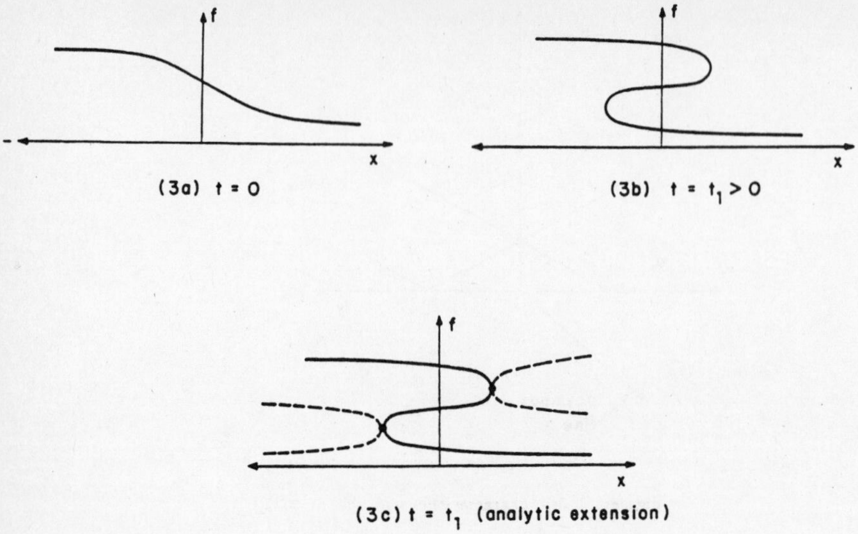

Figure 3. Solution of Burgers equation. Plot of u vs. x for (3a) initial data and (3b) time after multi-valuedness develops. (3c) Analytic extension of solution in (3b). Solution f is complex on dashed curves. Turning points in (3b) and (3c) are branch points for analytic extension.

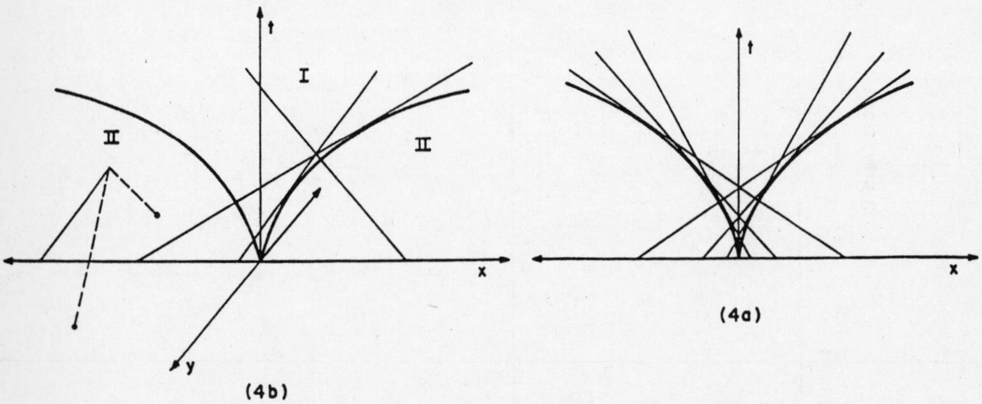

Figure 4. Characteristics in x vs. t for multi-valued solutions: solid lines are real characteristics, dashed lines are complex characteristics, dark curves are envelopes of characteristics. (4a) shows real characteristics and their envelopes. (4b) shows real and complex characteristics coming to points in two different regions.

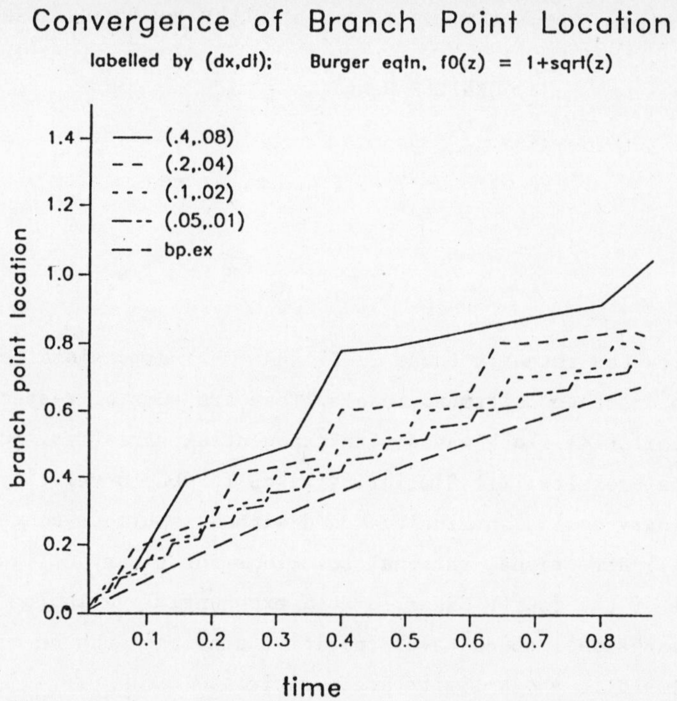

Figure 5. Location of branch point in x vs. t for Burgers equation (4.1) and numerical approximations of (4.1). Initial data is $f_0(z) = 1 + \sqrt{z}$. Note $h^{1/2}$ convergence rate.

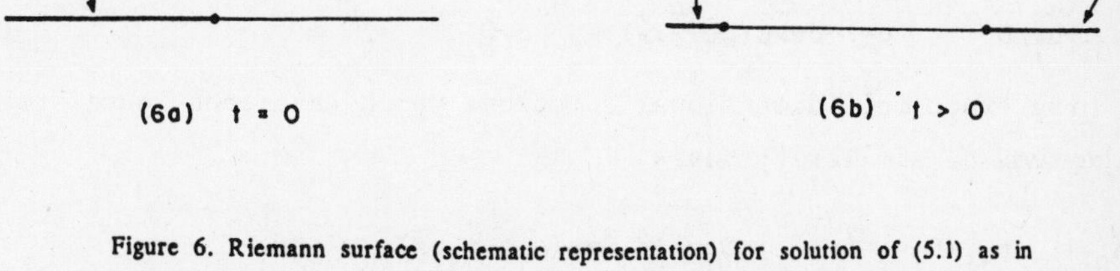

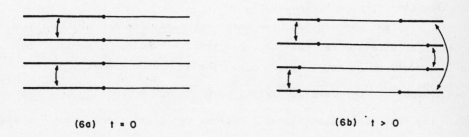

Figure 6. Riemann surface (schematic representation) for solution of (5.1) as in Theorem 2. (6a) is initial 4 sheeted surface with single branch point. (6b) is surface for $t > 0$ with separated branch points.

EXACT EXPONENTIAL TYPE SOLUTIONS TO THE DISCRETE BOLTZMANN MODELS

CORNILLE Henri

Service de Physique de Saclay

91191 Gif-sur-Yvette Cedex, France

ABSTRACT

We review the recently found (1+1) and (2+1)-dimensional exact solutions to the discrete Boltzmann models. They are sums of respectively two and three similarity shock waves with exponential variables. Further we prove original results: (i) Similarity waves for the 6-velocity planar model with ternary collisions included and without spurious conservation law (ii) (1+1)-dimensional rational solutions for binary collisions and denominators of the fourth order in two exponential variables (iii) For the (2+1)-dimensional shock waves ,positive densities can be constructed once the asymptotic shock-limits are positive.

1. INTRODUCTION

For the discrete Boltzmann models (DBM)[1] the velocity can only take discrete values $\vec{v}_i, |\vec{v}_i|=1$, $i=0,..2p-1$. To each $\vec{v}_i$ is associated a density $N_i(\vec{x},t)$ ($\vec{x}$ spatial coordinate $x_0, x_1 ..$). There exist p couples of opposite velocities $\vec{v}_i+\vec{v}_{i+p}=0$, $i=0,..p-1$.

The simplest exponential type exact solutions are the similarity shock waves first obtained for the cubic $6v_i$ Broadwell model[1], second[2] for some $2v_i$ models while now they are known for all current models

$$N_i=n_{0i}+n_i/D \qquad D=1+d\exp(\rho t+\vec{\gamma}.\vec{x}) \qquad d>0 \tag{1.1}$$

and the exact multidimensional solutions which have been found are simply the sums of similarity waves

$$N_i=n_{0i}+\sum_{j=1}^{J} n_{ji}/D_j \qquad D_j=1+d_j\exp(\rho_j t+\vec{\gamma}_j.\vec{x}) \qquad d_j>0 \tag{1.2}$$

with J=2 for the (1+1)-dimensional solutions and J=3 for the (2+1) ones.

The first (1+1)-dimensional exact solutions obtained were periodic[3] solutions for the Carleman $2v_i$ model. Periodic and other solutions were determined[4] for the Broadwell model[1], the general $2v_i$ Illner model[1], the $4v_i$ and $6v_i$ planar models[1] and periodic[5] solutions for the Cabannes[1] model. Different classes of (1+1)-dimensional solutions are known[3-4-5-6] (i) periodic spatial solutions nonpropagating with time $N_i=n_{0i}+2\mathrm{Re}(n_i/(1+d_i\exp(\rho_R t+i\vec{\gamma}_I.\vec{x}))$, (ii) shock waves sums of two real similarity components, (iii) solutions relaxing towards nonuniform Maxwellians $N_i=n_{0i}+n_i/(1+d_1\exp(\vec{\gamma}.\vec{x}))+p_i/(1+d_2\exp(\rho t))$, $\rho>0$, (iv) periodic spatial solutions propagating with t $N_i=n_{0i}+2\mathrm{Re}(n_i/(1+d_i\exp(\rho_R t+i(\vec{\gamma}_I.\vec{x}+\rho_I t))$ which can lead to sound-waves, (v) inclusion of simple ternary collisions which do not drop spurious conservation laws, (vi) specular reflection boundary condition at a wall.

For the (2+1)-dimensional exact solutions[7], three models have been investigated: the $4v_i$ planar model, the $6v_i$ cubic Broadwell model and as a generalization an hypercubic p-dimensional model which reduces to the previous ones for p=2,3. Three classes of solutions are known; (i) solutions relaxing toward nonuniform Maxwellians, (ii) shock waves sums of three real similarity waves (iii) semi-periodic solutions which are sums of (1+1)-dimensional periodic solutions plus a third similarity shock wave component.

The equations for the 2p planar models (Gatignol[1]) are:

$$(\partial_t+\vec{v}_i.\partial_{\vec{x}})N_i=(\partial_t-\vec{v}_i.\partial_{\vec{x}})N_{i+p}=-(p-1)N_iN_{i+p}+\sum_{\substack{k\neq i\\ k=0}}^{p-1}N_kN_{k+p}=\mathrm{Col}_i \tag{1.3}$$

with $\sum_i \mathrm{Col}_i=0$. There exist p+1 (>3 for p⩾3) linear relations and only three physical conservation laws in a two-dimensional space $\vec{x}=x_0,x_1$. Introduction of ternary collisions Col_T in addition to binary collisions Col_B leads to cubic nonlinearities instead of quadratic ones and the exact exponential type similarity waves have square-root branch point[6]

$$N_i=n_{0i}+n_i/D^{1/2} \qquad D=1+d\exp(\rho t+\vec{\gamma}.\vec{x}) \qquad d>0 \tag{1.4}$$

For models with only one independent Col_B a recipe was proposed by Harris[1] Gatignol[1] and Platkowski[8] which is to include ternary collisions of the type $\mathrm{Col}_T=(1+\tau M)\mathrm{Col}_B$, $M=\Sigma N_i$ being the total mass and (1+1)-dimensional solutions were obtained. However for the elimination of the spurious conservation law of the $6v_i$ hexagonal model, it was suggested(Hardy Pomeau[1], Gatignol[1-9]) to introduce another cubic term $\mu\mathrm{Col}_{T'}$ (μ being a constant) corresponding to $2\pi/3$ angles between the colliding particles. Due to a recent work may be no (1+1)-dimensional solutions exist[10]. In section 2 we present the unidimensional(not (1+1)-dimensional) similarity waves(1.1-4) for Col_B, $(1+\tau M)Col_B$, $(1+\tau M)Col_B$ and $\mu Col_{T'}$. The H Theorem requires $1+\tau M\geq 0$, $\mu>0$ and we prove that positive densities satisfying these two conditions exist.

The fact that the exact solutions found are sums of similarity solutions could appear strange because the DBM equations are nonlinear. This property does not hold for the continuous Boltzmann equation (CBE).For instance the popular Mott-Smith shock waves , sums of two Maxwellians, are not solutions of the CBE, only few moments satisfy the CBE(which represents another kind of discrete model). The linear conservation laws which are obtained after integration over the velocity space are not direct part of the CBE for the distribution functions. While the rational similarity solutions with an exponential variable u have denominators $(1+u)^{-2}$,those with two variables are of the type $(1+u_1+u_2)^{-2}$ and they do not correspond to sums of similarity waves for the CBE.Further, contrary to the DBM, inhomogeneous solutions relaxing towards absolute Maxwellians, have not been found without the introduction of external forces or sources[11]. What is so different in the DBM ? Let us write the p-dimensional hypercubic DBM equations

$$(\partial_t+\partial_{x_i})N_i=(\partial_t-\partial_{x_i})N_i=Col_i \qquad \sum_i Col_i=0 \tag{1.5}$$

with Col_i written down in (1.3). In both (1.3-5) models, among the 2p equations, we note that p+1(without spurious conservation law for (1.5)) which are linear are directly included into the DBM, leaving only p-1 nonlinear equations(However for the (1.3) models if we include all cubic, quartic[9] .. collisions necessary for the killing of unwanted linear relations, only 3 will remain).Introducing sums of similarity waves, first each component must be solution (which means relations between $n_{0i}, n_{ji}, \vec{\gamma}_j, \rho_j$ for j fixed). Second the sum being also a solution(no constraints from the linear relations) new relations, mixing j and j' arise:

$$-(p-1)(n_{ji}n_{j'i+p}+n_{j'i}n_{ji+p})+\sum_{k\neq i}(n_{jk}n_{j'k+p}+n_{j'k}n_{jk+p})=0 \qquad j\neq j' \tag{1.6}$$

If there exist sufficient parameters then (1.6) can be satisfied[4-6] and the sum can be a solution. It is clear that when J increases too much then the number of relations will exceed the number of parameters and such solutions do not exist anymore. For the hypercubic p-dimensional model, from a counting argument, we find $P=2p+J(3p+1)$ parameters and $R=2J+(p-1)(J^2+5J+2)/2$ relations. Seeking to construct (3+1)-dimensional solutions with $J=4$, we find $P=R=46$ for the Broadwell model and $P=60<R=65$ for the 4-dimensional one. For $J=3$ we get $R=13p-7\leq P=11p+3$ for $p\leq 5$.

In order to emphasize the importance of the linear relations for the selection of the possible rational (1+1)-dimensional solutions, in section3 we start with denominators of the type $D=1+\Sigma u_i+du_1u_2+\Sigma d_{3i}u_i^2u_j+\Sigma d_{4i}u_i^3u_j$, $u_i=d_i\exp(\gamma_i x+\rho_i t)$ and assume the existence of two independent linear differential relations. We prove that only sums of two similarity waves can exist. We show also that the only possible similarity waves are those of (1.1).

(2+1)-dimensional shock waves have been determined for the hypercubic $p\geq 2$ model(Notice that the previous counting argument is not rigorous in the sense that identities, reducing the number of relations, can occur). The total mass $M=\Sigma N_i$ has the same analytical structure $M=m_0+\Sigma m_j/D_j$ as the densities N_i (1.2). The main difficulty is the proof of the existence of positive N_i. In the arbitrary parameters space, from which we reconstruct all parameters, we must determine a domain leading to positive N_i. It is in fact sufficient to prove that at t=0 the asymptotic shock-limits in the coordinate plane are positive. For similarity waves, it is obvious that positive upstream and downstream shock-limits $n_{0i}, n_{0i}+n_i$ lead to $N_i>0$. The same property holds for (1+1) and (2+1)-dimensional shock waves solutions, provided the d_j in D_j satisfy well-defined bounds. In section 4 we consider functions of the type $Q(y_1,y_2)=q_0+\Sigma q_j/(1+d_j\exp y_j)$, $j=1,2,3$, $y_3=\tau_1y_1+\tau_2y_2$(which could be either N_i or M). We assume that the six asymptotic shock-limits(linear combination of the q_j, q_0) in the y_1,y_2 plane are positive. Then it is proved, provided the d_j satisfy well-defined bounds, that $Q>0$ in the whole plane.

2. SIMILARITY SHOCK WAVES FOR THE $6v_i$ PLANAR MODEL

We consider the hexagonal $6v_i$ model in one spatial dimension with ternary collisions(Gatignol[(1)]) included or not. For the ternary collisions we include both Col_T the collisions when two of the three particles have opposite velocities (parameter τ) and μCol_T, those with $2\Pi/3$ angles among the incoming and outgoing particles. We choose $\vec{v}_0, \vec{v}_3$ along the positive,negative x-axis. The four independent densities $N_0, N_3, N_1=N_5, N_2=N_4$ satisfy:

$$L_0N_0=Col_T+\mu Col_{T'} \quad L_3N_3=Col_T-\mu Col_{T'}, \quad L_3N_3-4L_1N_1=3L_0N_0, \quad L_0N_0-4L_2N_2=3L_3N_3$$

$$Col_T=(1+\tau M)Col_B \quad Col_B=N_1N_2-N_0N_3 \quad Col_{T'}=N_3N_1^2-N_0N_2^2$$

$$L_0=\partial_t+\partial_x \quad L_3=\partial_t-\partial_x \quad L_1=\partial_t+\partial_x/2 \quad L_2=\partial_t-\partial_x/2 \tag{2.1}$$

We have replaced $L_1N_1=-Col_T/2-\mu Col_{T'}$, $L_2N_2=-Col_T/2+\mu Col_{T'}$ by the two linear differential relations which are equivalent to the two conservations of mass and momentum. If $\mu=0$, then $L_0N_0=L_3N_3$ gives an additional spurious conservation law. In the binary $\tau=\mu=0$ and ternary cases $\mu=0$ $\tau\neq0$ and $\mu\neq0$ $\tau\neq0$ the exact similarity waves for N_i and $M=\Sigma N_i$ are:

$$N_i^B=n_{0i}+n_i/D \quad M^B=m_0+m/D \quad m=\Sigma n_i \quad m_0=\Sigma n_{0i}$$

$$N_i^T=n_{0i}+n_i/\sqrt{D} \quad M^T=m_0+m/\sqrt{D} \quad D=1+\exp\gamma u \quad u=x+ct \tag{2.2}$$

We omit the upperscript $n_{0i}^T, n_{0i}^B, .. \gamma^T, \gamma^B$ when it is not necessary. For the H-Theorem with $H=\Sigma N_i \log N_i$ we find: $\partial_t H+\partial_x(..)=Col_T\log(N_1N_2/N_0N_3)+\mu Col_{T'}\log(N_3N_1^2/N_0N_2^2)$ so that we require:

$$1+\tau M\geq0 \ (\text{ or } 1+\tau m_0\geq0 \text{ and } 1+\tau(m+m_0)\geq0\ ) \qquad \mu>0 \tag{2.3}$$

For the thickness $w=|m|/\mathrm{Max}|M_u|$ with $|M_u|\leq|\gamma|/4$ in the binary case, $|M_u|\leq|\gamma|/3\sqrt{3}$ in the ternary case, assuming the same shock-limits m_0, m_0+m and the same shock-speed c we find for w^T, w^B and their ratio R:

$$w^T=4|m/\gamma^B| \quad w^T=3\sqrt{3}|m/\gamma^T| \quad R=w^T/w^B=3\sqrt{3}/4\ |\gamma^B/\gamma^T| \tag{2.4}$$

We substitute (2.2) into (2.1) and successively study $\mu=\tau=0, \mu=0\ \tau\neq0, \mu\neq0\ \tau\neq0$.

2.1 Binary collisions alone (see also Ref.4)

We find 8 relations, 12 parameters, put $n_{00}=1$ and we can choose c,m,m_0 as the three arbitrary parameters from which we build-up the solutions. We define scaled parameters:

$$y=n_3/n_0 \quad \bar{n}_i=n_i/n_0 \quad i=1,2 \tag{2.5}$$

and construct the (c,m,m_0)-solutions with the 8 relations:

$$y=(c+1)/(c-1) \quad \bar{n}_1=-(c+1)/(2c+1) \quad \bar{n}_2=-(c+1)/(2c-1)$$

$$n_0=m(c-1)(2c+1)(2c-1)/6c \quad \rightarrow \quad n_3=yn_0 \quad n_i=\bar{n}_in_0 \tag{2.6}$$

$$n_{03}=n_{01}n_{02}=m_0-1-2n_{01}-2n_{02}=\zeta n_0(\bar{n}_1\bar{n}_2-y)+n_{01}\bar{n}_2+n_{02}\bar{n}_1-y \tag{2.7}$$

$$\gamma^B=n_0(\bar{n}_1\bar{n}_2-y)/(c+1) \tag{2.8}$$

Here $\zeta=1$ in (2.7). We notice that n_{01} (or n_{02}) are solutions of a quadratic equation with two possible choices.

2.2 Binary plus Ternary $\tau\neq 0$ $\mu=0$ alone collisions (see also Ref.6)

As in the binary case we still have the spurious conservation law. From (2.1) we see that the constant into Col_T must be zero. This leads to two classes of solutions $(1+\tau m_0)(n_{01}n_{02}-n_{00}n_{03})=0$ and for simplicity here we consider $1+\tau m_0\neq 0$. Comparing with the binary case we find one more relation:

$$\tau=1/(\zeta m-m_0) \qquad \zeta=\mp 1 \tag{2.9}$$

but with the additional τ parameter we can as above choose $n_{00}=1$ and c,m,m_0 as the arbitrary parameters. The relations (2.6) for the n_i are still valid, the only change in the n_{0i},n_i relations (2.7) is that now ζ can have two values ∓ 1. Finally with γ^B written in (2.8) we find:

$$\gamma^T=2m\tau\gamma^B \rightarrow R=3\sqrt{3}/(8|\tau m|) \tag{2.4'}$$

For the positivity, in both subsections 2.1-2, we must have $n_{0i}\geq 0$ $n_{0i}+n_i\geq 0 \rightarrow m_0\geq 0, m_0+m\geq 0$ and in 2.2 for the H-Theorem we add $1+\tau m_0\geq 0, 1+\tau(m+m_0)\geq 0$. From these constraints and (2.9-4') we find:[6] (i) If $\tau>0$ then necessarily $\zeta=1, m\geq m_0>0$, $R<3\sqrt{3}/8<1$, (ii) If $\tau<0$ and $\zeta=-1$, then necessarily $m>0$ $R>3\sqrt{3}/8$, (iii) If $\tau<0,\zeta=1$ and $-m_0<m<0$, then $R>3\sqrt{3}/4>1$.

As illustration we choose two numerical examples with $\zeta=1$(or the same n_{0i},n_i in both binary and ternary cases) and we consider two opposite τ signs: (i) $c=-0.91$, $m=2$, $m_0=7 \rightarrow n_{0i}=1$, 2.8, 0.08, 0.23, $i=0,1,2,3$ $n_i=1.62$, 1.8, 0.05, -0.07, $i=0,1,2,3$, $R=1.6$, $\tau=-0.2$; (ii) $c=-0.91$, $m=5.$, $m_0=4. \rightarrow n_{0i}$ =1., 1.03, 0.31, 0.32, $n_i=4.04$, 0.44, 0.13, -0.1 $\tau=1$ $R=0.13<1$ Gatignol[1]. We note in (2.2) the change of analytical structure for M when $\tau\rightarrow 0$ $(\zeta m-m_0\rightarrow\infty)$, for a discussion see Ref.4.

2.3 Binary plus Ternary collisions $\tau\neq 0$ $\mu\neq 0$, without spurious conservation law

From the vanishing of the constant of Col_T we still choose $1+\tau M\neq 0$. We find 14 parameters $n_{0i},n_i,m,m_0,c,\gamma,\tau,\mu$ subjected to 12 relations. We still have (2.9) that we explain: In the collisions terms,from the coefficients of $1/D, D^{-1/2}, D^{-3/2}$(the first is zero and for the equation $L_0N_0=..$ the two other are opposite) we find $(1+\tau m_0)A+\tau mB=0$, $\tau mA+(1+\tau m_0)B=0$. We deduce (2.9) and also $A=\zeta B$ or

$$n_1n_2-n_0n_3=\zeta(n_{03}n_0+n_{00}n_3-n_{01}n_2-n_{02}n_1) \qquad \zeta=\mp 1 \tag{2.9'}$$

We have one scaling parameter and <u>one arbitrary parameter</u> and <u>contrary to the above cases we cannot construct the microscopic N_i from the the macroscopic quantities</u> c,m,m_0. Besides (2.9-9') and the two m_0,m relations we find

$$c=(n_3+3n_0+2n_1)/(n_3-3n_0-4n_1) \qquad \gamma=4\tau m(n_1n_2-n_0n_3)/((c+1)n_0+(c-1)n_3)$$
$$4\mu/\gamma=((c+1)n_0-(c-1)n_3)/(n_1^2n_3-n_2^2n_0) \tag{2.10}$$

$$n_{01}n_{02}=n_{00}n_{03} \qquad n_{03}n_{01}^2=n_{00}n_{02}^2 \tag{2.11}$$

and four relations n_i,n_{0i}: (2.9') and

$$n_1^2n_{03}+2n_{01}n_1n_3=n_{00}n_2^2+2n_{02}n_0n_2 \qquad 8(n_0n_3+n_1n_2)+9(n_0n_2+n_1n_3)+n_2n_3+n_1n_0=0$$
$$n_1^2n_3-n_2^2n_0=n_{02}^2n_0+2n_{00}n_{02}n_2-n_{01}^2n_3-2n_{01}n_{03}n_1 \tag{2.12}$$

We introduce the same $y,\bar{n}_i$ parameters as in (2.5) and obtain:

$$y(8+9\bar{n}_1+\bar{n}_2)+8\bar{n}_1\bar{n}_2+9\bar{n}_2+\bar{n}_1=0, \qquad \bar{n}_2^2n_{00}+2\bar{n}_2n_{02}-2n_{01}\bar{n}_1y-\bar{n}_1^2n_{03}=0$$
$$n_0=\zeta(n_{03}+yn_{00}-n_{01}\bar{n}_2-n_{02}\bar{n}_1)/(\bar{n}_1\bar{n}_2-y) \tag{2.12'}$$
$$n_0^2=(n_{02}^2+2n_{00}n_{02}\bar{n}_2-n_{01}^2y-2n_{01}n_{03}y)/(\bar{n}_1^2y-\bar{n}_2^2)$$

First we notice from (2.9-10) that once the n_i,n_{0i} are determined then τ, c,γ,μ are deduced. Second from the n_{0i} relations alone (2.11) we can define $\bar{n}_{0i}=n_{0i}/n_{00}$, $\bar{m}_0=m_0/n_{00}$ and obtain: $\bar{n}_{02}=\bar{n}_{01}^3$, $\bar{n}_{03}=\bar{n}_{01}^4$, $\bar{m}_0=1+\bar{n}_{01}^4+2(\bar{n}_{01}+\bar{n}_{01}^3)$. n_{00} appears like a scaling factor for the n_{0i}, the same property holds for n_0(see (2.12)) and for all n_i. Consequently in the sequel we choose:

$$n_{00}=1, \quad n_{01}>0 \text{ arbitrary}, \quad n_{02}=n_{01}^3, \quad n_{03}=n_{01}^4 \tag{2.13a}$$

The substitution of y given by the first (2.12) relation into the second one leads to a cubic equation for $\bar{n}_2$ with $\bar{n}_1$-dependent coefficients. Similarly the elimination of n_0 into the two last (2.12) equations leads to another $\bar{n}_2$ polynomial. The solution of (2.12'-13a) is the following:

$$n_0=-\zeta,\ \bar{n}_1=n_{01},0=f(\bar{n}_2,\bar{n}_1)=\bar{n}_2^3+\bar{n}_2^2(9\bar{n}_1+8+2\bar{n}_1^3)+\bar{n}_2(18(\bar{n}_1^2+\bar{n}_1^4)+32\bar{n}_1^3-\bar{n}_1^6)+2\bar{n}_1^3\bar{n}_1^6(9\bar{n}_1+8) \tag{2.14}$$

Once the roots $\bar{n}_2$($\bar{n}_1$ or n_{01}-dependent) of f=0 have been obtained, then from the first (2.12') relation we find y and all n_i are known:

$$n_0=-\zeta \qquad n_1=-n_{01}\zeta \qquad n_2=-\bar{n}_2\zeta \qquad n_3=-y\zeta \tag{2.13a}$$

From the knowledge (2.13a-b) of the n_{0i}, n_i, we can deduce all other parameters: $m_0, m, c, \tau, \gamma, \mu$ (2.2-9-10).

We want to prove that there exist physically acceptable solutions with $N_i \geq 0$ (due to $n_{0i}>0$ in(2.13a) we must only prove $n_{0i}+n_i \geq 0$) satisfying the H-Theorem (or $1+\tau m_0>0, 1+\tau(m+m_0)\geq 0$). In the sequel we assume for the arbitrary parameter:

$$1/3<n_{01}=\bar{n}_1<1/2 \tag{2.15a}$$

Lemma 1: For the cubic $\bar{n}_2$ equation there exists one root $-\bar{n}_1^{-3}<\bar{n}_2<0$.
In $f(\bar{n}_2,\bar{n}_1)$ all $\bar{n}_1$-dependent coefficient of the $\bar{n}_2$ powers are positive, whence the roots $\bar{n}_2$ are negative. We easily find: $f(0,\bar{n}_1)>0$, $f(\bar{n}_2=-8,\bar{n}_1)>0$, $f(\bar{n}_2=-\bar{n}_1^{-3},\bar{n}_1)<0$. It follows that one $\bar{n}_2$ root is less than -8, another between -8 and $-\bar{n}_1^{-3}$ and the last one between $-\bar{n}_1^{-3}$ and 0. In the sequel we choose this $\bar{n}_2$ root of the cubic equation f=0 in (2.14)

$$-\bar{n}_1^{-3}<\bar{n}_2<0 \tag{2.15b}$$

Lemma 2: $-\bar{n}_1^{-4}<y<0$
From $n_0^2=1$ written down in (2.12') we get $2y\bar{n}_1^{-2}=(\bar{n}_2+\bar{n}_1^{-3})(\bar{n}_2-\bar{n}_1^{-3})+2\bar{n}_1^{-3}\bar{n}_2$. Applying Lemma 1 we have $\bar{n}_2+\bar{n}_1^{-3}>0, \bar{n}_2<0$ and then $y<0$. Adding $2\bar{n}_1^{-6}$ to both sides $2(y+\bar{n}_1^{-4})\bar{n}_1^{-2}=(\bar{n}_1^{-3}+\bar{n}_2)^2>0$.

Lemma 3: $\mu\gamma c>0$, $\bar{n}_1\bar{n}_2-y>0$, $\tau m\gamma c n_0<0, \zeta\mu\tau m>0$.
For the first result we have: $2\mu n_0^2/\gamma\bar{n}_1 c=(1+y(2+3\bar{n}_1)/\bar{n}_1)/(\bar{n}_2^{-2}-y\bar{n}_1^{-2})(y+3+2\bar{n}_1)$. The two factors in the denominator are positive while the numerator(Lemma 2) is larger than $1-\bar{n}_1^{-3}(2+3\bar{n}_1)>9/16$. For the second result we note $2(\bar{n}_1\bar{n}_2-y)=(\bar{n}_1^{-3}+\bar{n}_2)(\bar{n}_1^{-3}-\bar{n}_2)/\bar{n}_1^{-2}>0$ from Lemma 1. For the third one we get $2m\tau/cn_0\gamma=(4y-\bar{n}_1+3y\bar{n}_1)/(\bar{n}_1\bar{n}_2-y)(y+3+2\bar{n}_1)<0$ because the numerator is negative while the denominator is positive. The last result is a consequence of the previous ones and $n_0=-\zeta$

Lemma 4: $-\zeta m=1+y+2(\bar{n}_1+\bar{n}_2)>0$ is a consequence of $1+y>0$ and $\bar{n}_1+\bar{n}_2>0$.

Lemma 5: For the shock-speed we have $|c|<1$ and $c<1$.
We obviously get: $0<(y+3+2\bar{n}_1)/(-y+3+4\bar{n}_1)=-c<1$.

Theorem 1: For the solutions with $n_{01}=\bar{n}_1,\bar{n}_2$ satisfying (2.15a-b) and $\zeta=\mp 1$ then both $N_i\geq 0$, $1+\tau M\geq 0$, $\mu>0$, $\tau<0, c<0$, $\gamma<0$. If $\zeta=-1$ (or $n_0=1$) then $m>0$ while if $\zeta=1$ (or $n_0=-1$) then $m<0$.
We begin with the $\zeta=-1, n_0=1$ case and get: $m>0$ (Lemma 4), $\tau=-1/(m+m_0)<0$, $1+\tau(m+m_0)=0$, $1+\tau m_0=-\tau m>0$, $n_0+n_{00}=2$, $n_1+n_{01}=2n_{01}>0$, $n_2+n_{02}=\bar{n}_2+n_{01}^3>0$, $n_{03}+n_3=n_{01}^4+y>0$. From $\tau m<0$, $mcn_0\tau\gamma<0$ we get $c\gamma>0$ while $c\gamma\mu>0$ gives $\mu>0$.
We go on with the $\zeta=1, n_0=-1$ case and get: $m<0$, $\tau=1/(m-m_0)<0$, $1+\tau m_0=\tau m>0$, $1+\tau(m+m_0)>1+\tau m_0>0$, $n_0+n_{00}=0, n_1=-\bar{n}_1$, $n_{01}+n_1=0$, $n_2=-\bar{n}_2>0$, $n_{02}+n_2>0$, $n_3=-y>0$, $n_{03}+n_3>0$. From $\tau m>0$, $mcn_0\tau\gamma<0$ we get $c\gamma>0$ while $c\gamma\mu>0$ gives $\mu>0$.

In conclusion we have found two classes ($\zeta=\mp 1$ and $\bar{n}_1=n_{01},\bar{n}_2$ in the domain (2.15a-b)) of densities $N_i\geqslant 0$ satisfying the H-Theorem $1+\tau M\geq 0,\mu>0$. However for these solutions τ is negative and we adopt a phenomenological viewpoint. For the collision part $Col_T=(1+\tau M)Col_B$, here $0\leq 1+\tau M<1$. When these ternary collisions are included into the binary one then the additional factor is neither negative nor larger than 1. Of course the character of gain minus loss terms must be respected for Col_T, $\mu Col_{T'}$. This means that necessarily $\mu>0$, $1+\tau M\geq 0$ and these conditions are satisfied by the present similarity solutions. As illustration we present in fig.1 two examples $\zeta=\mp 1$ with the same $n_{01}=\bar{n}_1=0.5$ (a) $\zeta=-1$ $\rightarrow\tau=-0.23$, $\mu=0.93, c=-0.79$, $\gamma=-0.21;\bar{n}_2=-57\ 10^{-4}$, $y=-0.034$; $n_{0i}=1.,0.5,0.125,0.0625$, $n_i=1,0.5,-57\ 10^{-4},-0.034$ $i=0,1,2,3$. (b) $\zeta=1$ leading to the same parameters values except the n_i which have opposite values. In both cases $m_0=2.31$ while $m=1.95$ in case (a) and -1.95 in(b).

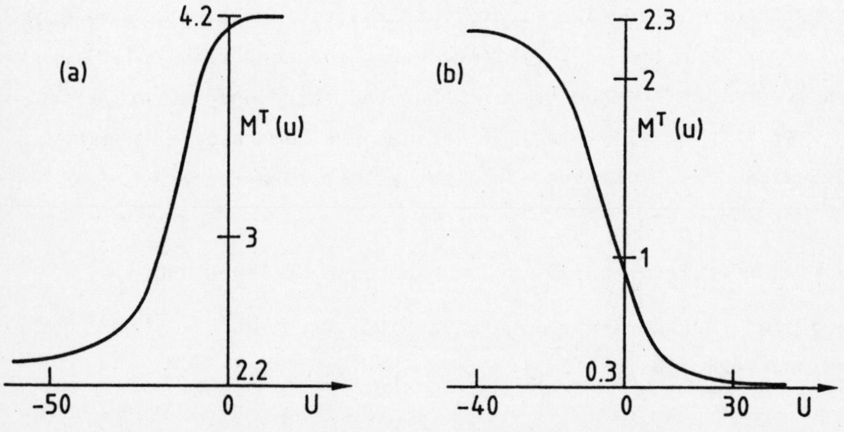

Fig. 1 - Shock profiles for $M^T(u=x+ct)$, $n_{01}=0.5$ (a) $\zeta=-1$ (b) $\zeta=1$

3. POSSIBLE EXPONENTIAL TYPE (1+1)-DIMENSIONAL RATIONAL SOLUTIONS FOR BINARY COLLISIONS

We want to show that the existence of two independent linear conservation laws give strong restrictions on the possible (1+1)-dimensional rational solutions with two exponential variables u_i

$$u_i=d_i\exp(\rho_i t+\gamma_i x),\ i=1,2 \qquad \rho_1\gamma_2\neq\rho_2\gamma_1 \tag{3.1}$$

Starting from a well-defined class of (1+1)-dimensional rational solutions we prove that the only possible ones are sums of similarity waves (two for the present class) taking into account only the existence of two linear relations. This result was previously obtained[(4)] for a smaller class. In one spatial dimension, with three independent densities called V,W,Z for both the $4v_i$ and the Broadwell model, they satisfy equations of the type:

$$V_t+V_x=-\alpha_1 W_t-\beta_1 W_x=-\alpha_2 Z_t-\beta_2 Z_x=\tau(Z^2-VW),\quad \alpha_1\beta_2\neq\alpha_2\beta_1,\quad \alpha_i\neq\beta_i \tag{3.2}$$

For the p=2,3 models (1.5) we have $x=x_0,\alpha_1=-1,\beta_1=1,\beta_2=0$. Further for the p=2 model we have $V=N_0,W=N_2,Z=N_1=N_3,\alpha_2=\tau=1$ while for the p=3 model $V=N_0,W=N_3$, $Z=N_1=N_2=N_4=N_5,\alpha_2=\tau=2$ and by a scaling $V\to V/\tau, W\to W/\tau\ldots$ we can choose $\tau=1$.

3.1 Exponential type similarity waves with the variable d $\exp\zeta$, $\zeta=\rho t+\gamma x$

Only here we take into account the nonlinearity in (3.2)and remark that V_ζ,W_ζ,Z_ζ are proportional. We can rewrite the densities and (3.2)

$$\begin{gathered}V=v_0+v/D,\ W=w_0+w/D,\ Z=z_0+z/D,\ D=D(\zeta)\quad D_\zeta+(a_0+a_1D+a_2D^2)/a=0,\quad a\neq0\\ a=v(\rho+\gamma)=-w(\alpha_1\rho+\beta_1\gamma)=-z(\alpha_2\rho+\beta_2\gamma),\ a_0=z^2-vw,\ a_1=2z_0z-v_0w-w_0v,\ a_2=z_0^2-v_0w_0\end{gathered} \tag{3.3}$$

D satisfies a Riccati equation with constant coefficients which is integrable. i) if $a_2\neq0$, putting $D=aE_\zeta/Ea_2$ we get a second order equation $a^2E_{\zeta\zeta}+aa_1E_\zeta+a_0a_2E=0$ with two different $\exp(\lambda_i\zeta)$ solutions for $A=a_1^2-4a_0a_2\neq0$. If A=0 the two solutions are $\exp\lambda\zeta$ and $\zeta\exp\lambda\zeta$ but for D remains only rational (linear in ζ) power type solutions that we disregard. If $A\neq0$ then D^{-1} can be written $c_1+c_2/(1+c_3\exp((\lambda_1-\lambda_2)\zeta)$, c_i being constants, which are of the exponential type with a rescaling of the arbitrary ζ. ii) If $a_2=0$ D is a constant plus an exponential. Choosing $a=-a_1=a_0$ we find for D:

$$D=1+d\exp(\zeta),\quad d>0 \tag{3.3'}$$

The determination of the parameters $v_0,w_0,z_0,v,w,z,\rho,\gamma$ leading to positive densities was done in Ref.4.

3.2 (1+1)-dimensional rational solutions $V(u_1,u_2)\ldots$

We require: i) when $u_j\to 0$, $u_i\neq 0$, the rational solutions reduce to the similarity ones: $V\to v_{00}+v/(1+u_i)$, $Z\to z_{00}+z/(1+u_i)$, $W\to w_{00}+w/(1+u_i)$; ii) when either u_i or u_j or both $\to\infty$ then the solutions are bounded (the degree of the numerator is at most equal to the denominator one); iii) only derivatives appearing into the linear relations of (3.2), if we write

$$V=v_{00}+\mathcal{V}/D,\quad W=w_{00}+\mathcal{W}/D,\quad Z=z_{00}+\mathcal{Z}/D,\quad D(u_1,u_2),\ \mathcal{V}(u_1,u_2)\ldots \tag{3.1'}$$

and redefine the constants $v_{00},\ldots$ we can always drop in $\mathcal{V},\mathcal{W},\mathcal{Z}$ a monomial $u_i^p u_j^q$ term present in in D. We want to find a criterion of the possible solutions at the linear level of the eqs.(3.2).

We study a class of possible rational solutions in (1+1) dimensions which must satisfy the two linear relations of (3.2)

$$D=1+\Sigma u_i+du_1u_2+\Sigma d_{3i}u_i^2u_j+\Sigma d_{4i}u_i^3u_j,\qquad \mathcal{V}=v_0+\Sigma v_{1i}u_i+vu_1u_2+\Sigma v_{3i}u_i^2u_j+\Sigma v_{4i}u_i^3u_j \tag{3.4}$$

and similar polynomials for $\mathcal{W},\mathcal{Z}$ with $(v_0,v,v_{1i},v_{3i},v_{4i})$ being replaced by $(w_0,w,w_{1i},w_{3i},w_{4i})$ and $(z_0,z,z_{1i},z_{3i},z_{4i})$. We sketch briefly partial results for a Theorem which is fully discussed in Ref.13.

Theorem 2: For the ansatz (3.4) satisfying the two linear conservation laws (3.2) the only possible solutions are the sums of two similarity waves

$$V=v_{00}+\sum_1^2 v_i/(1+u_i),\qquad W=w_{00}+\sum_1^2 w_i/(1+u_i),\qquad Z=z_{00}+\sum_1^2 z_i/(1+u_i) \tag{3.4'}$$

We rewrite the two (3.2) linear relations for $D,\mathcal{V},\mathcal{W},\mathcal{Z}$

$$\mathcal{V}(D_t+D_x)+\mathcal{Z}(\alpha_2D_t+\beta_2D_x)-D(\mathcal{V}_t+\mathcal{V}_x+\alpha_2\mathcal{Z}_t+\beta_2\mathcal{Z}_x)=0 \tag{3.4a}$$

and (3.4b) with $\mathcal{Z}\to\mathcal{W}$ $\alpha_2,\beta_2\to\alpha_1,\beta_1$. $(1+u_i)^{-1}$ and $(1+u_1^mu_2^n)$ being solutions of (3.4a-b) then sums of an arbitrary number of such similarity waves are solutions (for D in (3.4) this means sums of two $(1+u_i)^{-1}$). It is not trivial that they are the only ones. The complete proof requires a lot of algebra. Here, for simplicity, we assume $v_{31},v_{41}d_{31}$ and vd as $\neq 0$, but the general result holds[13] without such assumptions. The $u_i^pu_j^q$ terms being zero we check the compatibility of the $\mathcal{V}$,D parameters for the α_i,β_i i=1,2 values.

Lemma 6: $d_{4i}\neq 0$ is impossible. From $u_1^4u_2^4, u_i^5u_j^2, u_i^4u_j^3$ we obtain:

$$\sum_i(\alpha_2(\rho_i-\rho_j)+\beta_2(\gamma_i-\gamma_j))z_{4i}d_{4i}+(\rho_i-\rho_j+\gamma_i-\gamma_j)v_{4i}d_{4j}=0\quad i\neq j \tag{3.5a}$$

$$(\alpha_2\rho_i\ \beta_2\gamma_i)(z_{3i}d_{4i}-z_{4i}d_{3i})+(\rho_i+\gamma_i)(v_{3i}d_{4i}-v_{4i}d_{3i})=0 \tag{3.6a}$$

$$(\alpha_2(2\rho_i-\rho_j)+\beta_2(2\gamma_i-\gamma_j))(z_{3j}d_{4i}-z_{4i}d_{3j})+(2\rho_i-\rho_j+2\gamma_i-\gamma_j)(v_{3j}d_{4i}-v_{4i}d_{3j})=0 \qquad (3.7a)$$

and similar relations $z_{ki},\alpha_2,\beta_2 \to w_{ki},\alpha_1,\beta_1$ for the couple V,W. Due to (iii) we can choose v_{00},z_{00},w_{00} , not present in (3.5-6-7) such that $z_{42}=v_{42}=w_{42}=0$. We deduce if $d_{4i}\neq 0$:

$$z_{41}=v_{41}(\rho_2-\rho_1+\gamma_2-\gamma_1)/(\alpha_2(\rho_1-\rho_2)+\beta_2(\gamma_1-\gamma_2)),\quad z_{32}(\alpha_2\rho_2+\beta_2\gamma_2)=-v_{32}(\rho_2+\gamma_2)$$

$$z_{31}(\alpha_2\rho_1+\beta_2\gamma_1)=-v_{31}(\rho_1+\gamma_1)+(d_{31}v_{41}/d_{41})(\rho_1\gamma_2-\rho_2\gamma_1)(\alpha_2-\beta_2)/(\alpha_2(\rho_1-\rho_2)+\beta_2(\gamma_1-\gamma_2))$$

which substituted into (3.7a) gives a relation for the V,D parameters alone which must be true for the two couples of α_k,β_k values

$$-2v_{31}d_{41}/v_{41}d_{31}+1=(\alpha_k\rho_1+\beta_k\gamma_1)/(\alpha_k(\rho_1-\rho_2)+\beta_k(\gamma_1-\gamma_2)) \qquad k=1,2 \qquad (3.8)$$

The compatibility condition for the r.h.s.: $(\alpha_1\beta_2-\alpha_2\beta_1)(\rho_2\gamma_1-\rho_1\gamma_2)=0$ is not possible. In the sequel we assume $d_{4i}=v_{4i}=z_{4i}=0$.

Lemma 7: $d_{3i}\neq 0$ is impossible. From $u_1^3u_2^3, u_i^3u_j^2$ we get:

$$\sum_i(\alpha_2(\rho_j-\rho_i)+\beta_2(\gamma_j-\gamma_i))z_{3i}d_{3j}+(\rho_j-\rho_i+\gamma_j-\gamma_i)v_{3i}d_{3j}=0 \qquad j\neq i \qquad (3.9a)$$

$$(\alpha_2\rho_i+\beta_2\gamma_i)(zd_{3i}-dz_{3i})+(\rho_i+\gamma_i)(vd_{3i}-dv_{3i})=0 \qquad (3.9'a)$$

and the corresponding relations $z_{3i},z,\alpha_2,\beta_2 \to w_{3i},w,\alpha_1,\beta_1$ for W. Still from (iii) we can assume $z_{32}=v_{32}=w_{32}=0$ and obtain if $d_{3i}\neq 0$:

$$z_{31}=v_{31}(\rho_2-\rho_1+\gamma_2-\gamma_1)/(\alpha_2(\rho_1-\rho_2)+\beta_2(\gamma_1-\gamma_2)),\quad z=-v(\rho_2+\gamma_2)/(\alpha_2\rho_2+\beta_2\gamma_2)$$

which substituted into the i=1 eq(3.9'a) gives the V parameters alone relations

$$1-dv_{31}/vd_{31}=(\alpha_k\rho_1+\beta_k\gamma_1)/(\alpha_k\rho_2+\beta_k\gamma_2) \qquad k=1,2$$

with a compatibility condition $(\alpha_1\beta_2-\alpha_2\beta_1)(\rho_1\gamma_2-\rho_2\gamma_1)=0$ impossible. In the sequel we put $d_{3i}=v_{3i}=w_{3i}=z_{3i}=0$. Then from (iii) we can assume $z=v=w=0$ and the ansatz (3.4) is reduced to:

$$D=1+\Sigma u_i+du_1u_2\ ,\quad V=v_0+\Sigma v_{1i}u_i,\quad W=w_0+\Sigma w_{1i}u_i,\quad Z=z_0+\Sigma z_{1i}u_i \qquad (3.4')$$

Lemma 8: $d\neq 1$ is impossible, d=1 leads to $v_0=\Sigma v_{1i}$, $w_0=\Sigma w_{1i}$, $z_0=\Sigma z_{1i}$ and the solutions are sums of two similarity waves.

Assuming $d\neq 0$ we find from $u_i^2u_j, u_1u_2, u_i$:

$$d(z_{1i}(\alpha_2\rho_j+\beta_2\gamma_j)+v_{1i}(\rho_j+\gamma_j))=0 \qquad j\neq i \tag{3.10a}$$

def: $B_i=z_0(\alpha_2\rho_i+\beta_2\gamma_i)+v_0(\rho_i+\gamma_i)$ def: $A=\Sigma(\alpha_2(\rho_i-\rho_j)+\beta_2(\gamma_i-\gamma_j))z_{1i} +(\rho_i-\rho_j+\gamma_i-\gamma_j)v_{1i}$

$$\text{def: } B=\sum_i B_i \quad \rightarrow \quad dB=A \tag{3.11a}$$

$$\text{def: } A_i=z_{1i}(\alpha_2\rho_i+\beta_2\gamma_i)+v_{1i}(\rho_i+\gamma_i) \quad \rightarrow \quad B_i=A_i \tag{3.12a}$$

Substituting z_{1i} given by (3.10a) into both A_i and A defined in (3.10a-11a) we get:

$$A_i=(\alpha_2-\beta_2)v_{1i}(\rho_j\gamma_i-\rho_i\gamma_j)/(\alpha_2\rho_j-\beta_2\gamma_j) \qquad A=\sum_i A_i \tag{3.13}$$

Summing (3.12a) over i we find: $\Sigma B_i=B=\Sigma A_i=A=dB$, whence the result:

$$(d-1)B=0 \tag{3.11a'}$$

On the other hand (3.12a) with A_i written down in (3.13) gives two linear equations for z_0, v_0 easily solved and leading to:

$$v_0=\Sigma v_{1i} \qquad z_0=\Sigma z_{1i} \tag{3.12a'}$$

First we consider $d=1$ in (3.11a'). Then $D=(1+u_1)(1+u_2)$ and from (3.12a'):

$$V=v_{00}+\sum_i v_{1i}/(1+u_j),\ Z=..\quad W=..\qquad i\neq j \tag{3.14}$$

and the solutions are sums of two similarity waves. Second if $d\neq 1$ then $B=0$ is a linear combination of z_0, v_0. But $v_0=\Sigma v_{1i}$ and from (3.10a-12a') z_0 is also a linear combination of z_{1i}. We find:

$$v_{11}/v_{12}=(\alpha_k\rho_2+\beta_k\gamma_2)/(\alpha_k\rho_1+\beta_k\gamma_1)$$

and the difference for k=1,2 of the two r.h.s. still leads to an impossibility. It remains to study the case $d=0$ for which we can assume(see(iii)) that $z_{12}=v_{12}=w_{12}=0$. The three equations (3.11a-12a) become:

$$z_0(\alpha_2\rho_2+\beta_2\gamma_2)+v_0(\rho_2+\gamma_2)=0,\ z_{11}(\alpha_2(\rho_1-\rho_2)+\beta_2(\gamma_1-\gamma_2))+v_{11}(\rho_1-\rho_2+\gamma_1-\gamma_2)=0$$

$$(\alpha_2\rho_1+\beta_2\gamma_1)(z_0-z_{11})+(\rho_1+\gamma_1)(v_0-v_{11})0 \tag{3.15}$$

The first relation gives z_0, the second one z_{11} which substituted into the third leads to a relation between v_{11}, v_0

$v_{11}/v_0+1=-(\alpha_k\rho_1+\beta_k\gamma_1)/(\alpha_k\rho_2+\beta_k\gamma_2)$ $\qquad k=1,2$

still leading to an impossibility for k=1,2.

3.3 Models with only one conservation law

All the above impossibilities were coming from the existence of two conservation laws. What happens for the $2v_i$ models with only one conservation law? We write the equations with the most general collision term

$$N_{0t}+N_{0x}=-N_{3t}+N_{3x}=aN_0^2+bN_0N_3+cN_3^2 \qquad (3.16)$$

We recall[4] the results for the Illner model[1] with $a\leqslant 0$, $c\geqslant 0$, $b+a+c=0$. Starting with an ansatz of the type (3.4') $D=1+\Sigma u_i+du_1u_2\cdots$, using both the linear relation (3.16) and another relation proportional to D^{-2} in (3.16) we still obtain for the Illner model only d=1 or $D=(1+u_1)(1+u_2)$. (but positivity is violated for the (1+1) shock waves as well as for instance for the Carleman similarity shock wave[2-4]). (4)

We recall another result obtained in collaboration with T.T. Wu. Starting from the most general collision term (3.16) and seeking rational solutions with $D=1+u_1+u_2$ it was proved that the only possible case is provided with $a=c=0, b\neq 0$. The Ruijgrok Wu model[1] is of this type and it is completely soluble, which means that starting at t=0 with any rational function we can obtain a solution at ulterior time.

4. POSITIVE ASYMPTOTIC SHOCK-LIMITS AND POSITIVITY FOR MULTIDIMENSIONAL SHOCK WAVES

Let t=0 (or t_0 fixed) and consider a sum of p+1 real similarity shock waves (either densities N_i or total mass M) in a p-dimensional space:

$$Q=q_0+\sum_{j=1}^{p+1} q_j/D_j \qquad D_j=1+d_j\exp y_j \qquad d_j>0 \qquad (4.1)$$

with p independent variables y_j while y_{p+1} is a linear combination of the others.

For p=0(similarity waves), if the two shock-limits q_0, q_0+q_1 are positive then $Q(y_1)>0$ $\forall y_1$. For p=1(sums of two similarity waves) and if y_1, y_2 were independent variables we should have 4 shock-limits q_0, q_0+q_i $i=1,2, q_0+q_1+q_2$. With $y_2=\tau y_1$ exist only two limits; either $q_0, q_0+q_1+q_2$ if $\tau>0$ or q_0+q_i if $\tau<0$.

It was proved that if the two limits are positive, we can choose[4] appropriate d_j so that $Q>0\ \forall y_1$. More generally for $p\geq 1$, in the p-dimensional space, exist at least 2^p asymptotic limits and at most $2^{p+1}-2$ (4 and 6 for p=2 and 8,10,12,14 for p=3).

Let us consider (2+1)-dimensional shock waves, sums of three similarity waves, in a 2-dimensional coordinate space y_1,y_2 and $y_3=\tau_1 y_1+\tau_2 y_2$. The sums of the two first components give 4 limits q_0, q_0+q_i $i=1,2$, $q_0+q_1+q_2$ and we add the third component. Firstly we assume $\tau_1>0$ (or $\tau_1<0$) and $\tau_2=0$. Then we still have 4 limits, substituting q_1+q_3 to q_1 (or adding q_3 to the limits without q_1). It was proved[7] that if these 4 limits are positive, provided appropriate d_j were chosen, then $Q>0\ \forall y_1,y_2$. Secondly we assume $\tau_j\neq 0$ $j=1,2$, then in the y_1,y_2 plane the three lines $y_k=0$ $k=1,2,3$ define 6 domains which contain 6 asymptotic shock-limits. For instance if $\tau_j>0$ $j=1,2$ the limits are q_0,q_0+q_j $j=1,2$, $q_0+q_3+q_j$ $j=1,2$, Σq_k $k=0,1,2,3$; if $\tau_j<0$ $j=1,2$ the limits are: $q_0+q_j, q_0+q_k+q_j$ $j=1,2,3$ $k=1,2,3$ $j\neq k$ and still 6 limits for the two other cases $\tau_1\tau_2<0$. Among the 8 possible limits $q_0, q_0+q_j+q_k, \Sigma q_k, q_0+q_k$ always two of them are missing. We can rewrite Q with these 8 possible limits

$$Q\prod_j D_j=\sum_{k=0}^{k=3} q_k+\sum_i u_i\,(q_0+q_j+q_k)+\sum_{\substack{i,j\\ i\neq j}} u_i u_j (q_0+q_k)+q_0\prod_j u_j \tag{4.2}$$

If we assume that the six asymptotic limits are positive, due to the fact that two terms in (4.2)which are the missing limits, can be negative, then it is not obvious that the 6 positive terms could overcome the positivity problem in the whole y_1,y_2 plane. We prove this property, for simplicity, only in one $\tau_j>0$ of the four possible cases.

Theorem 3: If the six asymptotic limits q_0,q_0+q_i and $q_0+q_3+q_i$ $i=1,2$, $\Sigma_0^3 q_k$ are positive; if $Q=q_0+\Sigma_1^3 q_j/D_j$, $D_j=1+d_j\exp(y_j)$, $d_j>0$, $y_3=\Sigma\tau_i y_i$ $i=1,2$, $\tau_i>0$; if d_j satisfy well-defined bounds, then $Q(y_1,y_2)>0$ in the plane y_1,y_2.

Lemma 9: If $P=p_0+p/\Delta$, $0\leq\Delta^{-1}\leq 1$ then either $P>p_0$ if $p_0>0$ or $P>p+p_0$ if $p<0$.

In (4.2) the two limits not present into the 6 asymptotic limits and which can be negative are q_0+q_3 and $q_0+q_1+q_2$. For the three q_1,q_2,q_3 exist 8 possible signs and we apply Lemma 9 to all these cases: (i) if $q_1>0,q_2>0$, then

$Q>q_0+q_3/D_3$ and $Q>q_0+q_3$ if $q_3<0$(and we cannot conclude) whereas $Q>q_0>0$ if $q_3>0$; (ii) if $q_1q_2<0$ applying the Lemma 9 to the two possible signs of q_3 we always find a positive lower bound belonging to one of the 6 positive asymptotic limits; (iii) if $q_1<0, q_2<0$ then $Q>q_0+q_1+q_2+q_3/D_3>\Sigma q_k>0$ if $q_3<0$ but we cannot concluded if $q_3>0$. At this stage remain two cases: $q_1>0$ $q_2>0$ $q_3<0$ $q_0+q_3<0$ and $q_1<0$ $q_2<0$ $q_3>0$ $q_0+q_1+q_2<0$ for which the positivity of Q is unknown.

Lemma 10: If the assumptions of Theorem 3 hold and if $q_1>0$ $q_2>0$ $q_3<0$ $q_0+q_3<0$, provided the d_j satisfy well-defined bounds, then $Q>0$ $\forall y_1 y_2$.

We prove that in (4.2) the only negative term $(q_0+q_3)u_1u_2$ is dominates by positive ones. Let us take y_0 an arbitrary fixed number and consider, in the y_1,y_2 plane the four domains: (i) $y_i \geq y_0$ $i=1,2$ (ii) $y_i \leq y_0$ (iii) $y_1<y_0$ $y_2>y_0$ (iv) $y_1>y_0, y_2<y_0$.

First in domain (i) $u_3 \geq d_3 \exp(\tau_1+\tau_2)y_0$ and $u_1u_2(q_0+q_3+q_0u_3)>0$ if

$$d_3>(|q_0+q_3|/q_0)\ \exp-(\tau_1+\tau_2)y_0 \tag{4.3a}$$

Second in (ii) $u_1u_2 \leq d_1d_2\exp 2y_0$ and $\Sigma q_i+u_1u_2(q_0+q_3)>0$ if

$$d_1d_2< \Sigma q_i(\exp(-2y_0))\ /|q_0+q_3| \tag{4.3b}$$

Third in (iii) $u_1 \leq d_1\exp y_0$ and $u_2(\ q_0+q_1+q_3+u_1(q_0+q_3))>0$ if

$$d_1<(q_0+q_1+q_3)(\exp -y_0)/|q_0+q_3| \tag{4.3c}$$

Fourth in (iv) $u_2 \leq d_2\exp y_0$ and $u_1(q_0+q_2+q_3+u_2(q_0+q_3))>0$ if

$$d_2<(q_0+q_2+q_3)(\exp -y_0)/|q_0+q_3| \tag{4.3d}$$

Lemma 11: If the assumptions of Theorem 3 hold and if $q_1<0$ $q_2<0$ $q_3>0$ $q_0+q_1+q_2<0$, provided the d_j satisfy well-defined bounds, then $Q>0$ $\forall y_1,y_2$.

We prove that in (4.2) the only negative term $u_3(q_0+q_1+q_2)$ is dominated by positive ones in the y_1,y_2 plane. First in the domain (i) $u_1u_2 \geq d_1d_2\exp(2y_0)$ and $u_3(u_1u_2q_0+q_0+q_1+q_2)>0$ if

$$d_1d_2>|q_0+q_1+q_2|\,(q_0)^{-1}\exp -2y_0 \tag{4.4a}$$

Second in (ii) $u_3 \leq d_3\exp(\tau_1+\tau_2)y_0$ and $\Sigma q_i+u_3(q_0+q_1+q_2)>0$ if

$$d_3<(\Sigma q_i/|q_0+q_1+q_2|)\exp-(\tau_1+\tau_2)y_0 \tag{4.4b}$$

Third in (iii) $u_2 \geq d_2 \exp y_0$ and $u_3(u_2(q_0+q_1)+q_0+q_1+q_2)>0$ if

$$d_2>(|q_0+q_1+q_2|/(q_0+q_1))\exp -y_0 \tag{4.4c}$$

Fourth in (iv) $u_1 \not\geq d_1 \exp y_0$ and $u_3(u_1(q_0+q_2)+q_0+q_1+q_2)>0$ if

$$d_1>(|q_0+q_1+q_2|/(q_0+q_2)) \tag{4.4d}$$

For the other τ_1,τ_2 signs cases, the positive 6 asymptotic shock-limits still control the positivity in the whole y_1,y_2 plane, provided the d_j satisfy well-defined lower or upper bounds (established like here for the $\tau_1>0,\tau_2>0$ case). In conclusion these shock-limits are the quantities which control the positivity and for the total mass they are the relevant macroscopic physical quantities.

REFERENCES

1. Broadwell J.E. Phys.Fluids 7 1243(1964); Harris S. Phys.Fluids 9 1328 (1966); Hardy J. Pomeau Y. J.Math.Phys. 13 1042(1972); Gatignol R."Lect. Notes in Phys." 36 1042(1975),TTSP 16 837(1987); Cabannes H."Lect. Notes" Berkeley(1980), TTSP 16 809(1987); Illner R. Math.Meth.Appl.Sc. 1 187(1979) Ruijgrook Th. Wu T.T. PhysicaA 113 401(1982)

2. Platkoski T. J.Mec.Theo.Appl.4 555(1985); Dudek G. Nonnenmacher T.F. PhysicaA135 167(1987)

3. Bobylev A.V. Math.Congr.Warsaw Book ABstract B29(1983); Wick J.Math.Met. Appl.Sci. 6 515(1984); Piechor K. quoted in Platkoski paper Ref.2

4. Cornille H. J.Phys.A20 1973(1987),J.Math.Phys.28 1567(1987),Phys.Lett.A 125 253(1987),J.Stat.Phys.48 789(1987)

5. Cabannes H. Tiem D.M. CRAS304 29(1987), Complex Systems 1 574(1987)

6. Cornille H."Advances Electronics Electron Phys." Inverse Prob.Ed.Sabatier p.481(1987), CRASS304 1091(1987), J.Math.Phys.29 1667(1987), Lett.Math. Phys. 16 245(1988)

7. Cornille H. J.Phys.A20 L1063(1987), "Some Topics in Inv.Prob."Ed.Sabatier World Scient. p.101(1988), J.Stat.Phys.52 897(1988), Proc.XVI Rar.Gas. Dyn.Pasadena(1988), Saclay Preprints SPhT/88-113-158 TTSP and J.Mat.Ph.1989 Proc. Lattice Gas,Discrete Kinet.Theor. Torino 1988

8. Platkoski T. Bull.Pol.Acad.Sc.32 247(1984), Mec.Res.Comm.11 201(1984)

9. Coulouvrat F. Gatignol R. CRAS306 392(1988)

10. Longo E. Monaco R. Proc.XVI Rar.Gas Dyn. Pasadena(1988)

11. Cornille H. J.Stat.Phys.45 615(1986)

12. Ince E.L." Ordinary Diff.Eq." p.311 (Denver 1956)

13. Cornille H. Saclay Preprint Spht/89-27

SEMICONDUCTOR MODELLING VIA THE BOLTZMANN EQUATION [1]

P. DEGOND [2]

F. GUYOT-DELAURENS [2]

F.J. MUSTIELES [2]

F. NIER [2]

Abstract: This paper is devoted to the presentation of a new numerical method for the simulation of the Boltzmann Transport Equation of semiconductors, the weighted particle method.

In this paper, we will describe the kinetic model of the Boltzmann Equation and present the numerical method that we propose; we will provide numerical results of simulations in two distinct cases: first an homogeneous problem and then an inhomogeneous one, where one has to solve a coupled Boltzmann-Poisson system.

---

[1] This work has been partially supported by the "Centre National d'Etudes des Télécommunications", under grant n° 878B087 LAB/ICM/TOH , by the "Direction des Etudes et Recherches Techniques", under grant n° 87/283 and by the CNRS, under the "ATP-Mathématiques et Informatique".

[2] Centre de Mathématiques Appliquées, Unité Associée CNRS n° 756; Ecole Polytechnique; 91128 Palaiseau Cedex; France.

1. INTRODUCTION

Most of the numerical simulations of semiconductor devices use the drift-diffusion model [4,13]. This model is based upon Ohm's law (for drift) and Fick's law (for diffusion) and states that the drift term of the average velocity of the carriers is written :

$$v(E) = \mu(E) \, . \, E$$

where μ is a field dependent mobility ; this relation is obtained at equilibrium, as a consequence of the balance between the free acceleration of the carriers and their diffusion by the defects of the crystal lattice. The time needed for this equilibrium to be reached is the momentum relaxation time (mean time between collisions) so that Ohm's law is valid as long as this relaxation time is shorter than the time needed for the carriers to cross the device. But, for a 1 micron long Gallium Arsenide device these times are about the same and therefore Ohm's law is not valid. For instance, in such devices, some carriers have almost collisionless (or ballistic) flights and thus the average velocity can be higher than Ohm's law predicted value [4,5,6]. In fact, the drift-diffusion model does not take into account the main features of transport in submicronic devices [7] : the presence of ballistic carriers, the large proportion of high velocity ("hot") carriers and the large gradients of carrier density and temperature.

In order to account for these features, a modification of the drift-diffusion model has been proposed by many authors [5,7,8] : it is the hydrodynamic model. It consists of conservation equations for the mass, momentum and energy and is deduced from the Boltzmann Transport Equation by the moment method under the assumption that the distribution function of the carriers is a drifted Maxwellian distribution. Scattering processes are accounted for by empirically defined relaxation times for momentum and energy at the right hand side of these conservation equations; some other modifications [6,9,10] enable to include some thermal effects. Nevertheless, this model fails to describe accurately the ballistic and hot electrons effects, and is based upon a highly questionable assumption. No satisfactory hydrodynamic model is available yet.

The kinetic model (the Boltzmann Equation) then seems to give the most accurate description of the physics attainable by numerical computations. In this paper, we will recall the main features of the semiconductor Boltzmann equation and describe the weighted particle method that we use for its simulation ; we will provide numerical results of such simulations in two different cases. First, the choice of an homogeneous test problem enabled us to validate our method, focusing on the description of the collision operator [1,2] ; we will present here the results obtained for an infinite sample of semiconductor imbedded in a constant electric field, with a physically realistic collision operator ; we will also describe the case of a bidimensional electron gas near the interface of an heterojunction. Second, we adopted a simplified collision operator and concentrated on the resolution of a coupled Boltzmann-Poisson system [14] ; we will present the different Poisson solvers that we compared. For more details about the model, the numerical method and its mathematical analysis, we refer the reader to [1,2,3] ; for a more physical description of the kinetic model, see for example [4,11,12].

2. THE SEMICONDUCTOR BOLTZMANN EQUATION

With most generality, the semiconductor Boltzmann equation describes the two species of carriers, electrons and holes by means of two distribution functions, respectively $f_n(x,k,t)$ and $f_p(x,k,t)$, which represent the number of electrons or holes with position x and wave-vector k at time t, averaged over a small volume dxdk in the phase space ; the positions x are in a bounded set Ω (the device geometry) while the wave-vectors belong to the first Brillouin zone B.

The evolution of these distribution functions is governed by the system of Boltzmann equations:

$$\partial_t f_n + v_n(k).\nabla_x f_n - (q/\hbar)E.\nabla_k f_n = Q_n(f_n) + S_n(f_n,f_p)$$
$$\partial_t f_p + v_p(k).\nabla_x f_p + (q/\hbar)E.\nabla_k f_p = Q_p(f_p) + S_p(f_p,f_n)$$

for $x \in \Omega$, $k \in B$, $t \geq 0$; q is the elementary charge and $\hbar$ is the reduced Planck's constant ; $v_n(k)$ and $v_p(k)$ are known vector fields giving for each specie the velocity associated with a wave-vector k. E(x,t) is the electric field given via Poisson's equation ; $Q_n(f_n)$ and $Q_p(f_p)$ are the scattering operators for electrons and holes while $S_n(f_n,f_p)$ and $S_p(f_p,f_n)$ stand for recombination terms between these two kind or carriers.

From now on and throughout the paper we will suppose that the electrons are the only charge carriers in the device. This simplification enables us to write the Boltzmann equation of the electron distribution function f in the following way:

$$\partial_t f + v(k).\nabla_x f - (q/\hbar)E(x,t).\nabla_k f = Q(f)(x,k,t)$$
$$x \in \Omega \subset R^3 , k \in B \subset R^3 , t \geq 0 \qquad (1)$$

the electric field E(x,t) is related to the electron density n(x,t) via Poisson's equation:

$$E(x,t) = - \nabla\phi (x,t) \qquad (2)$$

$$- \Delta\phi(x,t) = \frac{q}{\varepsilon} (n_D(x) - n(x,t)) \qquad (3)$$

$$n(x,t) = \int f(x,k,t) \rho_s dk \qquad (4)$$

ε and ρ_s are respectively the permittivity of the material and the density of states in the k-space ; n_D is a given doping profile.

The relation v(k) is a known vector field written :

$$v(k) = \frac{1}{\hbar} \nabla_k \varepsilon(k) \tag{5}$$

where $\varepsilon(k)$ is the energy versus wave-vector relation, called the band diagram. We will suppose (unless specified) that the band diagram is parabolic :

$$\varepsilon(k) = \frac{\hbar^2 k^2}{2\, m^*} \ . \tag{6}$$

m^* is the effective mass of the electron in the crystal lattice. Let us notice that realistic band diagrams are far more complicated (different non parabolic valleys in Gallium Arsenide for example).

These equations must be supplemented by an initial condition for (1) :

$$f(x,k,0) = f_0(x,k)$$

and by boundary conditions for (1) and (3) :

$$\begin{aligned} f(0,k,t) &= g_0(k,t) \qquad \text{for } v(k) \geq 0 \\ f(L,k,t) &= g_L(k,t) \qquad \text{for } v(k) \leq 0 \\ \phi(0,t) &= \phi_0(t) \ ; \ \phi(L,t) = \phi_L(t) \end{aligned}$$

with f_0, g_0, g_L, ϕ_0 and ϕ_L suitably given.

The integral scattering operator Q(f) is written :

$$Q(f)(x,k,t) = \tag{7}$$

$$= \int_B \left[S(x,k',k)f(x,k',t)(1 - f(x,k,t)) - S(x,k,k')f(x,k,t)(1 - f(x,k',t))\right] dk'$$

S(x,k',k) are known transition rates depending upon the physical nature of the involved scattering process. The (1 - f) factors originate from Pauli's exclusion principle and make Q a non linear operator. For some examples of transitions rates, we refer the reader to [2,4,23], but we want to stress here that most of the transition rates may be written in the form :

$$S(x,k,k') = \sum \phi(x,k,k')\ \delta(\ \varepsilon(k') - \varepsilon(k) \pm \hbar\omega_p\ ) \tag{8}$$

the sum is to be taken over + and - respectively standing for the emission and the absorption of a phonon of energy $\hbar\omega_p$ by an electron, and over all the possible scattering mechanisms. The Delta function accounts for the conservation of the energy of the electron/phonon system during the collision. The function $\phi(x,k,k')$ depends upon the scattering mechanism and whether it is an absorption or emission process.

A simplified collision operator, which we will refer to as the relaxation time model is obtained by neglecting the (1 - f) factors in (7) and by setting :

$$S(x,k,k') = \frac{1}{\tau} M(k')$$

where τ is a constant relaxation time, and M(k) is a Maxwellian distribution associated with the lattice thermal velocity v_{th} :

$$M(k) = \frac{1}{2v_{th}} \exp\left(-\frac{v^2}{2v_{th}^2}\right) \tag{9}$$

Thus the collision operator in the relaxation time model is written :

$$Q(f) = -\frac{1}{\tau}(f - nM) \tag{10}$$

The coupled system of Boltzmann equation (1) and Poisson's equation (2,3,4) is non linear and has a large number of degrees of freedom. Moreover, Poisson's equation drives collision damped plasma oscillations, the frequency of which is a very limiting time scale. In the practical situations, the doping profile n_D is a quickly varying function, and leads to a high dimension stiff problem ; this will raise many numerical difficulties.

3. THE NUMERICAL METHOD : GENERAL PRESENTATION

The most widely spread numerical method for solving the semiconductor Boltzmann equation is the Monte-Carlo method (cf [4] and references therein), although other methods have been tried in particular geometries (cf Reed's method [16]) or for particular collision operators (see the recent method developped by Baranger [6] or Kuivalainen and Lindberg [17]). The Monte-Carlo method is extremely noisy. Thus, except in particular geometries, the affordable number of particles is not sufficient to get a sharp resolution of the distribution function, by stastical average. Only moments of the distribution function such as the current or energy densities can be recovered with a sufficiently sharp resolution, and only through time averages which do not allow the description of the transient regimes. The new methods and algorithms which we will describe in this paper are somehow derived from the Monte-Carlo method but are expected to have a better behaviour.

The weighted particle method was first introduced by G.H. Cottet, S. Mas-Gallic and P.A. Raviart [18,19], for viscous perturbations of the incompressible Euler equation. Then, the method was adapted to the treatment of collision terms in kinetic equations [20]. Its first application to the semiconductor Boltzmann equation has been done in [1, 2] and an error analysis relevant to this particular physical context has been performed in [3].

In this deterministic particle method, the particles move along the characteristics of the convective (first order differential) part of the equation, while the collision term is taken into account through the variation of the weights of the particles. The collision integral is evaluated by a discrete quadrature where the particles themselves play the role of quadrature points.

The weighted particle method is based upon the following approximation of the distribution function by a sum of Delta measures:

$$f(x,k,t) \approx f_h(x,k,t) = \sum_{i=1}^{N} \omega_i f_i(t)\, \delta(x - x_i(t)) \otimes \delta(k - k_i(t)) \qquad (11)$$

$x_i(t)$, $k_i(t)$, $f_i(t)$ and $\omega_i(t)$ are respectively the position, the wave-vector, the weight and the control volume of the particle i ; they evolve in time according to:

$$\frac{dx_i}{dt} = v(k_i) \qquad x_i(0) = x_i^0 \qquad (12\text{ a})$$

$$\frac{dk_i}{dt} = -\frac{q}{\hbar} E_i(t) \qquad k_i(0) = k_i^0 \qquad (12\text{ b})$$

$$\frac{df_i}{dt} = Q_i(t) \qquad f_i(0) = f_i^0 \qquad (13)$$

$$\omega_i(t) = \omega_i^0 \qquad (14)$$

where $E_i(t)$ and $Q_i(t)$ are the approximations of the electric field and of the collision operator acting on the i-th particle. The initial x_i^0, k_i^0, f_i^0 and ω_i^0 are chosen so that:

$$f_0(x,k) \approx \sum_{i=1}^{N} \omega_i^0 f_i^0\, \delta(x - x_i^0) \otimes \delta(k - k_i^0) \qquad (15)$$

The time differential system (12, 13) can be solved by any classical scheme. In our computations, we used the order 2 Adams Bashforth scheme.

To define $Q_i(t)$, we introduce a cut-off function $\zeta_\alpha(x)$ such that:

$$\zeta_\alpha(x) = \zeta\left(\frac{x}{\alpha}\right) \quad ; \quad \zeta(x) = \zeta(-x) \quad ; \quad \int \zeta(x)dx = 1 \qquad (16)$$

where ζ is a compactly supported function. We write (omitting the t-dependence of x_i, k_i, f) :

$$Q(f)\,(x_i,k_i) = \tag{17}$$

$$= \int \left[\; S(x_i,k',k_i)\, f(x_i,k')\,(1 - f(x_i,k_i)) - S(x_i,k_i,k')\, f(x_i,k_i)\,(1 - f(x_i,k')) \;\right] dk'$$

$$= \int \left[\; S(x',k',k_i)\, f(x',k')\,(1 - f(x_i,k_i)) - S(x_i,k_i,k')\, f(x_i,k_i)\,(1 - f(x',k')) \;\right] \delta(x' - x_i)\, dx'dk'$$

$$\approx \int \left[\; S(x',k',k_i)\, f(x',k')\,(1 - f(x_i,k_i)) - S(x_i,k_i,k')\, f(x_i,k_i)\,(1 - f(x',k')) \;\right] \zeta_\alpha(x' - x_i)\, dx'dk'$$

$$\approx \sum_{j=1}^{N} \left[\; S(x_j,k_j,k_i)\, f(x_j,k_j)\,(1 - f(x_i,k_i)) - S(x_i,k_i,k_j)\, f(x_i,k_i)\,(1 - f(x_j,k_j)) \;\right] \zeta_\alpha(x_j - x_i)\, \omega_j$$

Therefore we let (again omitting the t-dependence of k_i, x_i, and f_i) :

$$Q_i(t) = \sum_{j=1}^{N} [S(x_j,k_j,k_i)\, f_j\,(1 - f_i) - S(x_i,k_i,k_j)\, f_i\,(1 - f_j)]\, \zeta_\alpha(x_j - x_i)\, \omega_j \tag{18}$$

for our computations, we use the "hat function" W_2 as cut-off function ζ ; W_2 is a B-spline, defined after :

$$W_p = \chi^{*p} \tag{19}$$

where χ is the characteristic function of $[-1/2,1/2]$

We now turn to the computation of $E_i(t)$. The electric field in the device is the sum of an external field originating from the bias voltage and an internal field (of zero mean value over the device) resulting from the mutual Coulomb interaction between the charged particles. In an homogeneous case, we neglect this last term so that we write :

$$E_i(t) = E(t) \tag{20}$$

where E is the exterior field the device is imbedded in. On the other hand, in an inhomogeneous case, the mutual interaction between particles is taken into account and $E_i(t)$ is known from the resolution of Poisson's equation. This computation will be detailed in paragraph 5 where we present an inhomogeneous test problem.

4. THE HOMOGENEOUS CASE

Throughout this paragraph, we will suppose that the electric field does not depend upon the position variable x. The exterior electric field denoted by E will be supposed to have a constant direction and norm. The invariance of the Boltzmann equation under rotations allows us to chose the field axis as a reference axis, and to use an axisymmetric geometry. Therefore we write:

$$\frac{\partial f}{\partial t}(k,t) - \frac{qE}{\hbar}\frac{\partial f}{\partial k_1}(k,t) = Q(f) \qquad (21)$$

$k = (k_1, k_2) \in R \times R^+$; $k_1 = k.u$; $k_2 = | k - k_1 u |$ where u denotes a unitary vector in the direction of E.

The dependence upon the x variable has vanished as well as the coupled Poisson's equation. Equations (11), (12b), (13), (14), (15), (18) and (20) describe the weighted particle method in this particular case.

Figure 1 shows a comparison between simulation results obtained by this method and Monte-Carlo ones taken from [22]. The simulations delt with bulk Gallium Arsenide at temperature T = 77 K imbedded in a constant electric field E = 10 kV/cm. The band diagram of Gallium Arsenide is described by a standard three valley model, and the integral scattering operator was chosen according to (7), (8) with 40 different physical scattering mechanisms [23]. The curves of mean velocity, mean energy and density show a very good agreement between our results and the Monte-Carlo ones. Figure 2 displays three-dimensional views of the distribution function at different times of the simulation. We think such views can be of interest to get a better insight of the physics involved. For more details about this homogeneous test problem, we refer the reader to [1, 2, 23].

The next example of homogeneous model we present deals with a bidimensional electron gas near an heterojunction interface. At such a junction between two different materials of different band structures, a sharp potential well appears, resulting in discrete quantum states [24, 25]. The propagation of the electrons tranverse to the interface can be described by discrete energy states or "minibands" of the band diagram, while their movement parallel to the interface is classical. These structures are expected to give rise to higher velocities than bulk structures and are a deep concern of many ultrafast device conceptors. Our aim was to describe the transport properties of such an electron gas when a constant electric field is applied parallel to the interface. For a detailed presentation of physical models for heterojunctions, see [25, 26] ; the application of the weighted particle method to this case is detailed in [14].

The main features of the simulation are the following. Firstly, we solve iteratively a coupled system of Schrödinger and Poisson equations to find the energy levels, the associated wave functions and the equilibrium partition function of the electrons in the different levels ; we then compute the "overlap integrals" which appear in the functions $\phi(k,k')$ of formula (8) and measure the probability of band to band transitions in terms of the corresponding wave functions. Hence, we know the energy and velocity versus wave-vector relations, the equilibrium distribution of the electrons and the collision rates (8) of the scattering operator. These data are then used to solve the Boltzmann part of the model.

Indeed, the electronic population of each miniband is described by a distribution function $f_n(k,t)$ where $k=(k_x,k_y)$ is the wave-vector of the electron parallel to the interface. These distribution functions are solutions of the coupled system of Boltzmann equations:

$$\frac{\partial f_n}{\partial t} + \frac{qE_{//}}{\hbar}\frac{\partial f_n}{\partial k_x} = \sum_{p=1}^{N} Q_{n,p}(f_n,f_p) \qquad 1 \leq n \leq N \qquad (22)$$

where $E_{//}$ denotes the constant electric field applied parallel to the interface and N is the number of modelled minibands ; $Q_{n,p}(f_n,f_p)$ is the scattering operator between the electrons of the two minibands of indexes n and p.

As the electrons are heated up by the electric field and change band because of the collisions, the discrete energy levels, the wave functions and the potential well shape are modified; thus these quantities should be updated during the simulation of the transient regime. We have not done this updating yet and the results we present were obtained with constant energy levels, wave functions and well shape.

Figure 3 shows results of a simulation at a temperature T = 77 K, with an electromotive field $E_{//}$ = 2 kV/cm ; three minibands were modelled. Only one scattering mechanism was taken into account : the polar optical interaction where the electrons absorb or emit a phonon of constant energy. The lowest energy band has most of the electrons but gets depleted after one picosecond as the field heats them up to higher energy bands.The non monotonous variation of the population of the minibands can be explained by the relative values of the energy at the bottom of each band and of the thresholds for the emission of one or more phonons. The average velocity (over all the particles of the three bands) shows a characteristic overshoot profile. These figures show that it would be of interest to account for the variation of the discrete energy levels, the wave functions (thus the transition rates) and the well shape during the transient regime, particularly as far as the stationary value of the mean velocity is concerned.

5. AN INHOMOGENEOUS CASE

In this paragraph, we will concentrate on the resolution of Poisson's equation coupled to Boltzmann equation for a one dimensional structure. No assumption is made concerning the electric field, and the mutual Coulomb interaction between charged particles will be taken into account. On the other hand, a simplified collision operator is used for all the displayed simulations : the time relaxation model (9), (10). For more details about the model, the different Poisson solvers we will compare and their numerical analysis, we refer to [15] and references therein.

The equations in this one dimensional case are the Boltzmann equation (1) and Poisson's equations (2), (3), (4). The doping profile $n_D(x)$ we chose is written :

$$\begin{aligned} n_D(x) &= N^+ \quad \text{for} \quad 0 \le x \le x_1 \\ &= N^- \quad \text{for} \quad x_1 \le x \le x_2 \\ &= N^+ \quad \text{for} \quad x_2 \le x \le L \end{aligned}$$

with $N^- = 2.\ 10^{15}\ cm^{-3}$ and $N^+ = 10^{18}\ cm^{-3}$; This choice enables us to compare our results with those obtained by Baranger in [6]. The behaviour of the device is dominated by the dynamics of the carriers in the N^- region [30], and thus numerical methods are required to give a precise description of this region, which is not easy; indeed, the large inhomogeneity ($N^+/N^- = 500$) leads to a stiff problem where the numerical errors on the electron density in the N^+ region are of the same order of magnitude as the density itself in the N^- region; moreover, if the trajectories are not accurately solved, fast particles may jump over the peak of the electrid field instead of "seeing" it.

Again, equations (11) to (18) describe the weighted particle method in this case, except for the computation of $E_i(t)$ (12 b) which we present now.

For the approximation of $E_i(t)$, we considered two methods : the classical "Particle in Cell" (PIC) method [27,28], and an exact computation of the field acting on each particle using the Green's function of Poisson's equation.

In the PIC method, one introduces a grid of equally spaced points $X_m = m\,\Delta x$, $m = 0\ldots M_x$ and an interpolation function $W(x)$; the approximation of the density at the grid points is obtained by an assignment procedure :

$$n_m = \sum_{j=1}^{N} \omega_j f_j(t) \frac{1}{\Delta x} W\left(\frac{x_j(t) - X_m}{\Delta x}\right) \tag{27}$$

then, one solves Poisson's equation by finite differences on the grid X_m and gets an approximation of the electric field at the grid points E_m; the field is interpolated at the location of the ith particle :

$$E_i(t) = \sum_{m=0}^{M_x} E_m W\left(\frac{x_i(t) - X_m}{\Delta x}\right) \tag{28}$$

in our computations, we chose W_2 (see 19) as a cut-off function.

The Green's function method relies on an exact representation of the mutual Coulomb interaction between particles. From a direct integration of (11) with respect to v, we get a particle representation of the charge density according to :

$$\rho(x,t) \approx \rho_h(x,t) = q\left[n_D(x) - \sum_{j=1}^{N} \omega_j f_j \delta(x - x_j(t)) \right] \tag{29}$$

We denote by Q(t) and U(t) the following quantities :

$$Q(t) = \int \rho_h(x,t)\, dx = q\left[\int n_D(x)\, dx - \sum_{j=1}^{N} \omega_j f_j \right]$$

$$U(t) = \phi_L(t) - \phi_0(t)$$

From (29), we have the exact representation of the electric field :

$$E_h(x,t) = \frac{q}{\varepsilon} [\int K(x,y) n_D(y) dy - \sum_{j=1}^{N} \omega_j\, f_j(t)\, K(x,x_j(t))] + \frac{1}{L} [\frac{Qx}{\varepsilon} - U(t)] \tag{30}$$

where K(x,y) is the kernel associated to the Green's function G(x,y) defined after :

$$K(x,y) = - \frac{\partial G}{\partial x}(x,y)$$

$$- \frac{\partial^2}{\partial x^2} G(x,y) = \delta(x-y) - \frac{1}{L} \quad ; \quad G(0,y) = G(L,y) = 0$$

Thus the electric field $E_i(t)$ acting on the particle i is given by :

$$E_i(t) = \frac{q}{\varepsilon} [\int K(x_i(t),y) n_D(y) dy - \sum_{j=1}^{N} \omega_j\, f_j(t)\, K(x_i(t),x_j(t))] + \frac{1}{L} [\frac{Qx_i(t)}{\varepsilon} - U(t)] \tag{31}$$

The convolution term and the last bracket are easy to compute. In order to compute the summation term, we order the particles by increasing positions (using a fast sorting algorithm of cost O(NlogN)) and then use recursion formulae to compute the electric field $E_i(t)$ in order N operations. For a bidimensional problem, a fast algorithm recently derived by L.Greengard and Rokhlin [29], based upon multipoles expansions, would lead to a computation of cost O(N).

A regularization procedure can be used, replacing K(x,y) by a smoothed kernel in (31) :

$$K^{\alpha}(x,y) = (K(.,y) *_x \zeta_{\alpha})(x) \tag{32}$$

where ξ_{α} is a cut-off function satisfying (16). Since this procedure was not found to greatly improve the numerical results, but raises the computer cost, we do not derive it here and we refer the reader to [15] for more details.

We now turn to the comparison of the numerical results obtained with these two Poisson solvers. Figure 4 shows comparisons between the electric field computed by the PIC Poisson solver and the Green's function method, in a zero applied voltage case (thus the stationary equilibrium solution is known and provides an easy benchmark) ; for the PIC solver, we represent the field E_m at the grid points while in the Green's function method, we represent the "particle field" $E_i(t)$ (31). The Green's function method gives a precise description of the electric field in the lowest doped region and of the peak ; on the other hand, there are large fluctuations in the highly doped region, generated by the randomness of the positions of the particles in the phase space. We think these fluctuations have a very weak influence on the particles trajectories because they are completely random and of zero mean value. The PIC method also gives an oscillatory electric field in the N^+ region, but these oscillations are coherent, with a mesh dependant wavelength : they influence the dynamics of the particles because they create bunches of particles oscillating between close mesh cells, and thus generate mesh oscillations. Furthermore, the PIC method smears the peak of the field and reduces its value by a half. For a required quality of results, the Green's function method is performed in the same computer time as the PIC method, but requires half the number of particles of the PIC method ; this will turn to an advantage when using a real collision operator because its computation cost grows at least linearly with the number or particles. For all these reasons, we think that the Green's function method is preferable.

A second series of tests are concerned with an applied bias $U(t) = \phi_L(t) - \phi_0(t) = 0.47$ Volt, which corresponds to Baranger's simulations [6]. The initial datum is the equilibrium distribution function under zero applied voltage and thus the simulation mimics the transient evolution of the structure when a bias is suddenly applied at $t = 0$. The electric field is computed with the Green's function method. An important diagnostic in device simulations is the total current density $J(x,t)$ defined by :

$$J(x,t) = J_{part}(x,t) + \varepsilon \frac{\partial E}{\partial t}(x,t)$$

where J_{part} is the "particle current" :

$$J_{part}(x,t) = -q \int f(x,v,t)\, v\, dv$$

and $\partial E/\partial t$ is the displacement current. In this one dimensional structure, J does not depend on x because it is divergence free. Thus it is equal to its mean value over the device :

$$J(t) = \frac{1}{L}\int_0^L J(x,t)\, dx = \frac{1}{L}\iint f(x,v,t)v\, dx\, dv - \varepsilon \frac{\partial U}{\partial t}$$

In our example, U(t) does not depend on t. Therefore, we use the following approximation of the total current :

$$J(t) \approx \frac{1}{L} \sum_{i=1}^{N} \omega_i \, v_i(t) \, f_i(t)$$

Figure 5 displays the total current density J(t) versus time : it presents damped oscillations with initially large amplitudes which converge towards a stationary value approaching Baranger's one [6]. We think these oscillations may be related to the periodic boundary conditions that we use for the distribution function, but their frequency (intermediate between the plasma frequencies of the lowly and the highly doped region) may indicate a physical origin. We do not have a clear interpretation of these oscillations. On figure 5 are also displayed the electric field and the potential at time 1 ps for the same simulation ; the agreement between these results and reference [6] is very satisfactory : the relative difference on the electric field at the middle point of the N^- region is about 1%, and on the potential barrier, about 0.1%.

6. CONCLUSION

We have presented a new numerical method for solving the Boltzmann Transport Equation of semiconductors. Kinetic models seem to be the most suitable for the simulation of submicron devices because they are able to account for the characteristic features of transport in such devices. The numerical method we propose is a deterministic alternative to the Monte-Carlo method and may provide an interesting insight in the physics of electronic transport. It proved to be very satisfactory in the homogeneous axisymmetric test problem we presented, and enabled us to study the transport properties of an electron gas near an heterojunction interface. As for the inhomogeneous case, we investigated two methods for solving Poisson's equation self-consistently with the particle motion. The best method seems to be the Green's function method, but the PIC method nevertheless provides meaningfull results. Therefore, we think that the weighted particle method can help to get a precise description of the transient behaviour of real devices with moderate computer cost.

REFERENCES

[1] Niclot B., Degond. P., Poupaud F. *J. Comput. Phys.* 78 313 (1988)

[2] Degond P., Poupaud F., Niclot B., Guyot F. *Rapport interne n° 171, CMAP Ecole Polytechnique.* (1987)

[3] Degond P., Niclot B., "*Numerical Analysis of the weighted particle method applied to semiconductor Boltzmann equation*"; to be published

[4] L.Reggiani (ed) *Hot electron transport in semiconductor*; Topics in Applied Physics Series (1985); Springer, Berlin.

[5] Shur M.S., Eastman L.F. *IEEE Trans.* ED **26** (79) 1677

[6] Baranger H.U., Wilkins J.W. , *Phys. Rev. B* **30** (84) 7349; *Phys. Rev. B* **36** (87) 1487;Baranger H.U., PhD Thesis ,Cornell University, (1986)

[7] Rudan M., Odeh F., *Compel.* 5 (86) 149

[8] Cook R.K., Frey J., *Compel.* 1 (82) 65

[9] Cook R.K., Frey J.,*IEEE Trans.* ED 29 (82) 1970

[10] Bloetekjaer K., *IEEE Trans.* ED 17 (70) 38

[11] Rode D.L. in *Semiconductors and semimetals* **10**; Academic Press, NewYork (1975)

[12] Conwell E.M. in *Solid State Physics* **9** ; Academic Press, NewYork (1967)

[13] Selbehrerr S., *Analysis and Simulation of Semiconductor devices*; Springer Vierlag, Wien, NewYork (1984). Markovitch P.A. *The stationary Semiconductor Devuce Equations*; Springer Vierlag, Wien, NewYork (1986).

[14] Degond P., Guyot-Delaurens F., Nier F., Mustieles F.J., "*Simulation particulaire du transport bidimensionnel d'électrons parallèle à l'interface d'une hétérojonction*",manuscript, unpublished.

[15] Degond P; Guyot-Delaurens F., "*Particle Simulations of the semiconductor Boltzmann equation for one dimensional inhomogeneous structures*"; to be published.

[16] Rees H.D. *J. Phys. Chem. Solids* 30 643 (1969)

[17] Kuivainen P., Lidberg K., in *High Speed Electronics* edited by B.Küllback and H. Beneking; Springer Vierlag, NewYork (1986),p.40.

[18] Cottet G.H., Mas-Gallic S., to be published. Mas-Gallic S. and Raviart P.A., *Numer. Math.* 51 323 (1987).

[19] Degond P. and Mas-Gallic S., to appear in *Math Comput.*

[20] Mas-Gallic S. *Transp. Theory Stat. Phys.* **16** 855 (1987)
Mas-Gallic S., Poupaud F., *Transp. Theory Stat. Phys.* (to appear)

[21] Degond P; Mustieles F.J., manuscript; unpublished.

[22] Hesto P. Thèse, Université d'Orsay; (1984)

[23] Degond P., Mustieles F.J. *Le logiciel SPADES Documentation Scientifique*. (1988).

[24] Heiblum M., Eastman L.F., "*les électrons balistiques dans les semiconducteurs*", Pour la Science; Avril 1987.

[25] Vinter B.,*Appl. Phys. Lett.* 44 3 (1984)
Vinter B., *Appl.Phys. Lett.* 48 151 (1983)
Vinter B. in *Heterojunctions and Semiconductors Superlattices* edited by G.Allan, G. Bastard and al; Springer, Berlin (1986).

[26] M.Mouis, Thèse d'Etat, Université d'Orsay; 1988

[27] Hockney R.W., Eastwood J.W., *Computer Simulations using Particles*, McGraw Hill, NewYork; 1981.

[28] Birdsall C.K., Langdon , *Plasma Physics via Computer Simulations*, McGraw Hill, NewYork; 1985.

[29] Greengard L., Rokhlin V.J., *J. Comput. Phys.*73 325 (1987)

[30] Sze S.M., *Physics of Semiconductor Devices (2nd edition)*; John Wiley and Sons, New-York; 1981.

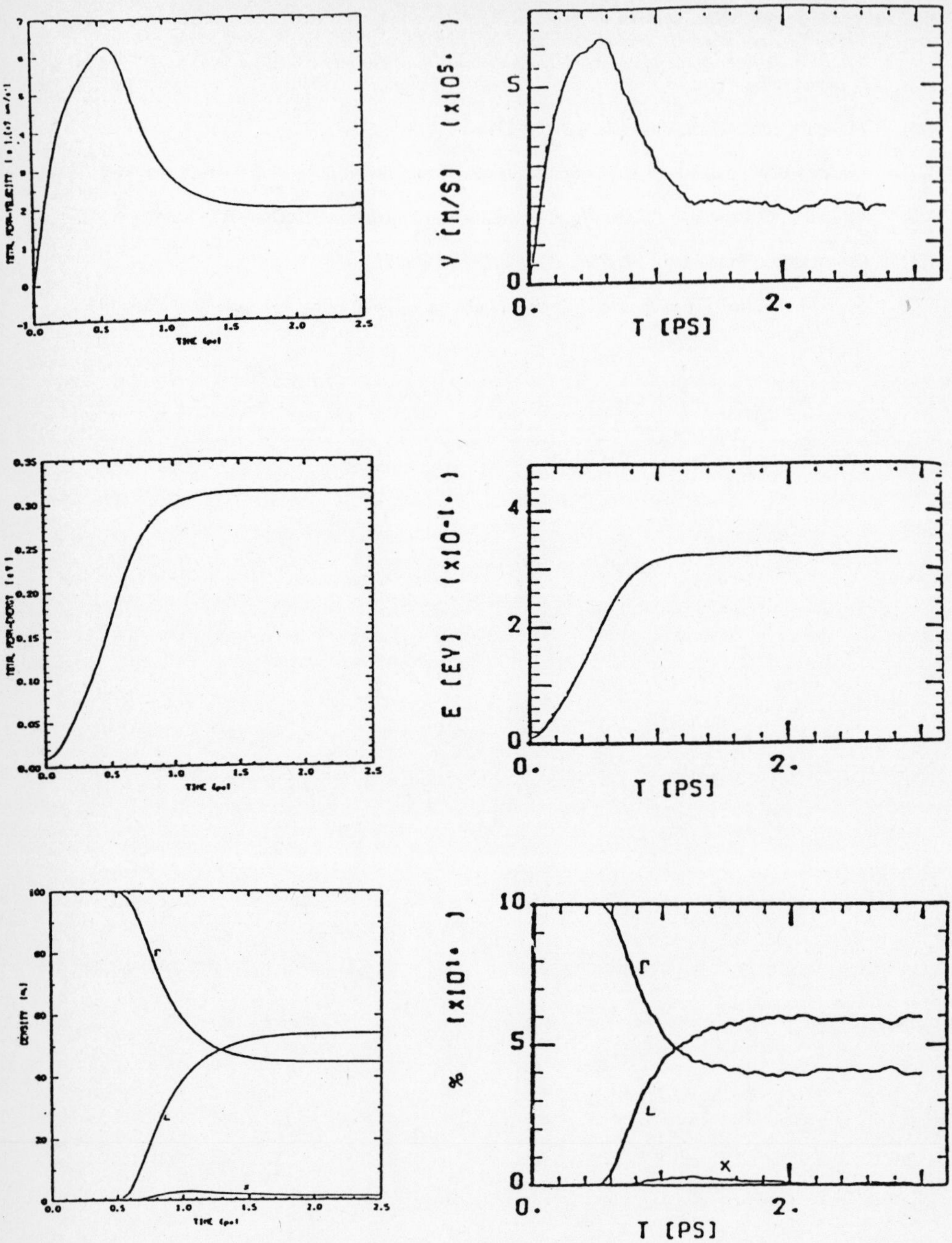

<u>Figure 1</u> mean total velocity (top, 10^7 cm/s), mean energy (middle, eV) and population of the three valleys Γ, L and X (bottom, percent of the total population) versus time (ps); left curves : our simulation ; right curves : Monte-Carlo simulation.

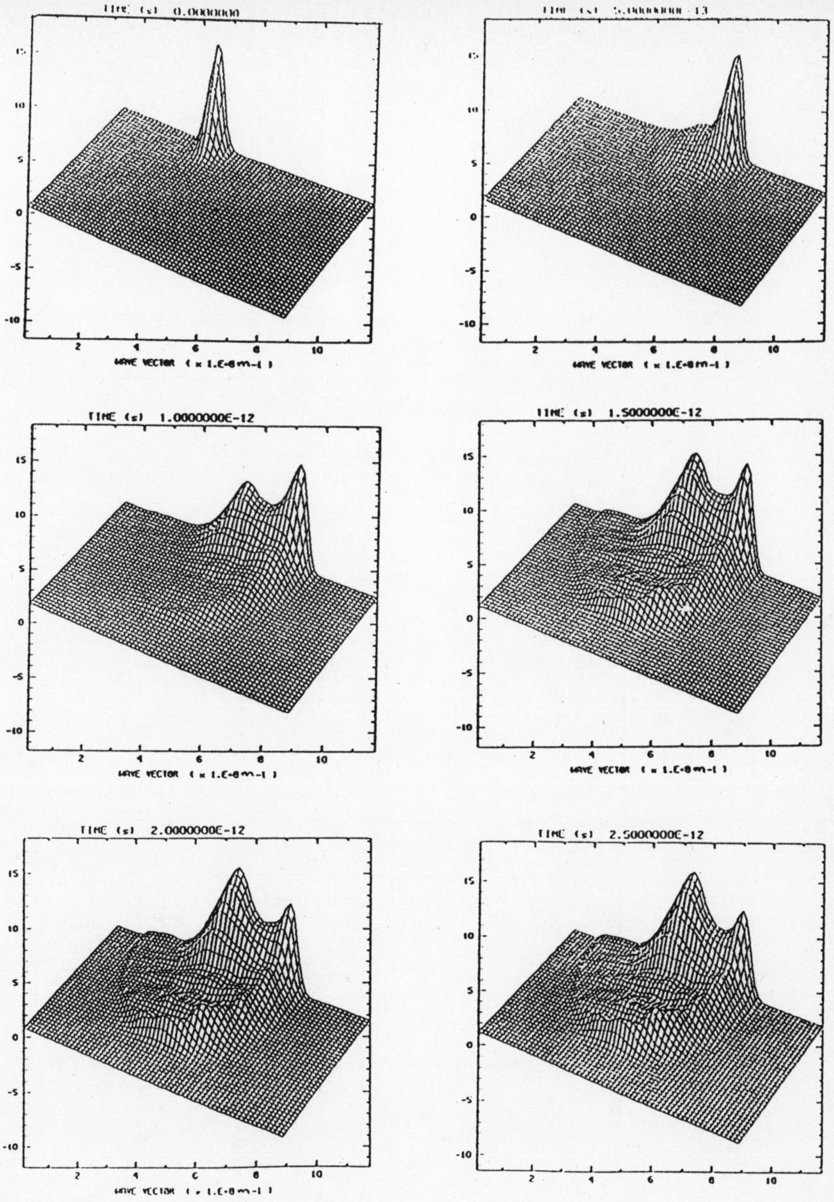

Figure 2 three dimensional views of the distribution function in the Γ valley at different times during the simulation.

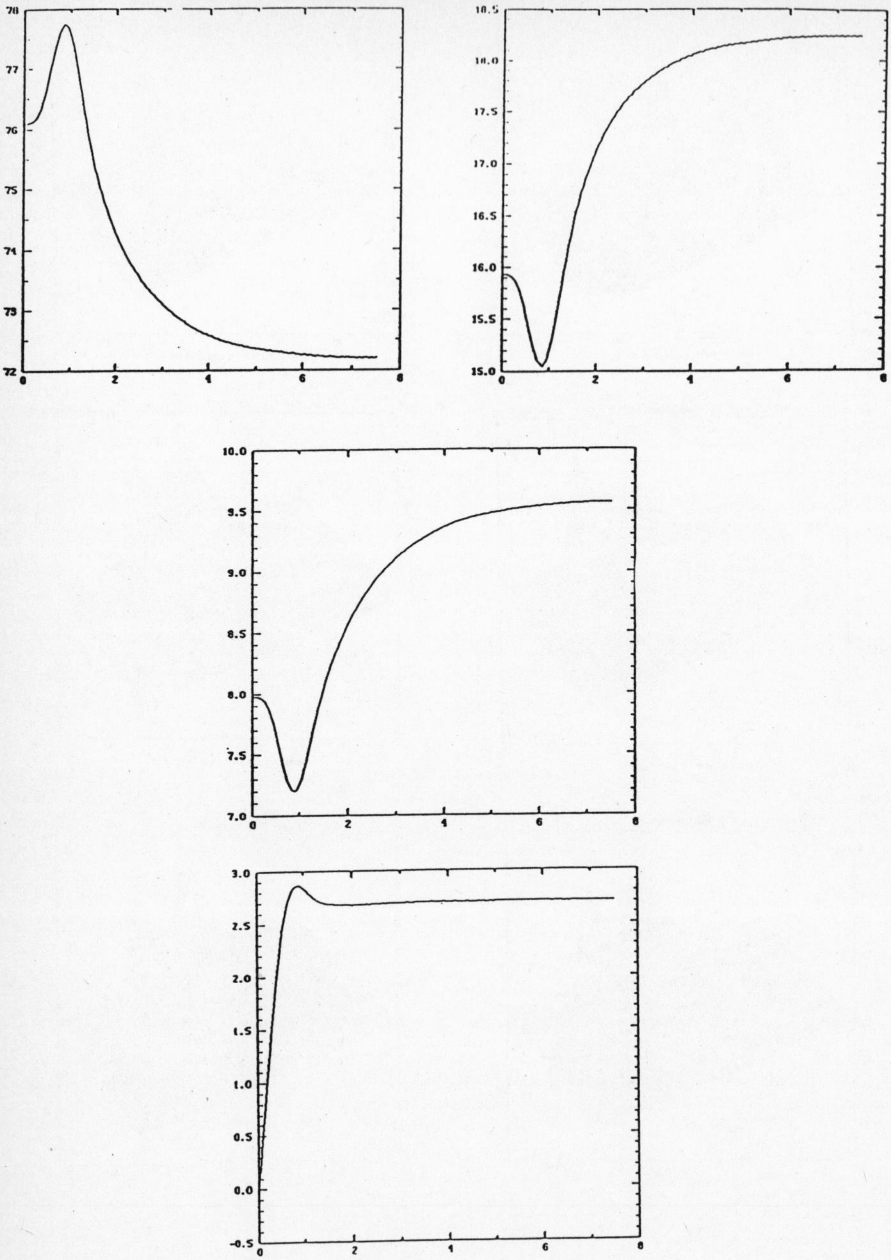

<u>Figure 3</u> population of each miniband in percent ot the total population versus time (ps); top left : lowest energy band; top right : first following band; middle : second following band; mean average velocity (10^7 cm/s) versus time (ps) (bottom)

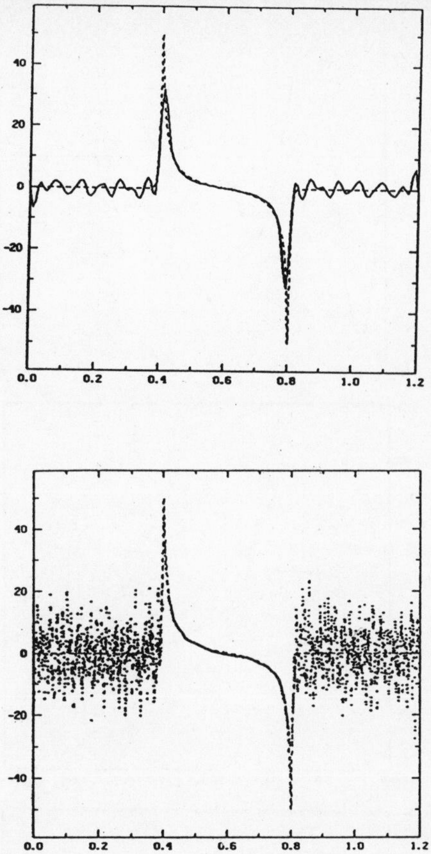

<u>Figure 4</u> electric field (in Kv/cm)as a function of distance (in μm) at time t = 1 ps, for 9000 particles; top figure : PIC method; bottom figure : Green's function method. The exact solution is shown in dash.

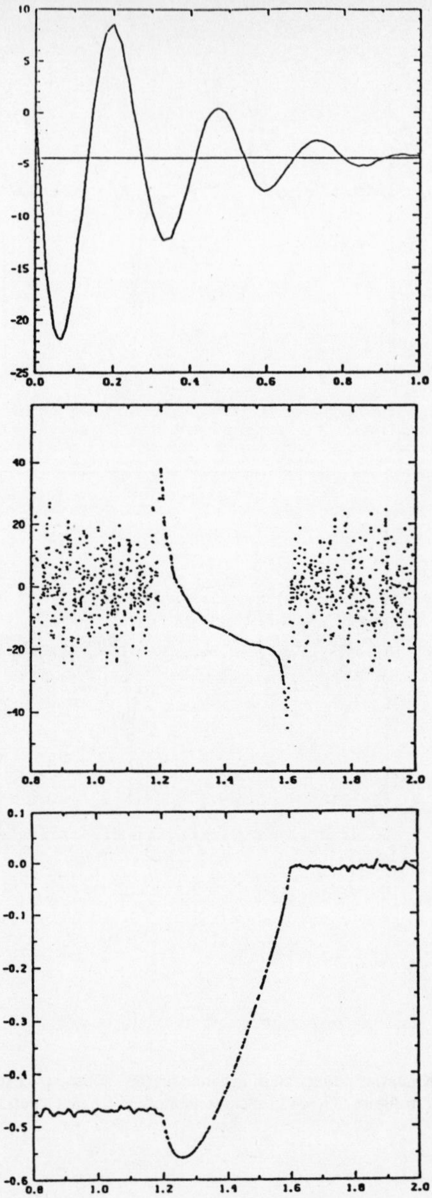

<u>Figure 5</u> top : current density (x 10^4 A/cm^2) as a function of time (in ps) ; the solid line gives the stationary value obtained by Baranger in [6] ; middle : electric field (in kV/cm) at time 1 ps as a function of the distance (in µm) ; bottom : electric potential (in Volt) at time 1 ps as a function of the distance (in µm) the length of the N^+ region is 1.2 µm and 30000 particles were used.

FUNCTIONAL-ANALYTIC TECHNIQUES IN THE STUDY OF TIME-DEPENDENT ELECTRON SWARMS IN WEAKLY IONIZED GASES

Giovanni Frosali
Dipartimento di Matematica, Università di Ancona
Via Brecce Bianche, I-60131 Ancona, Italy

ABSTRACT. The time-evolution of charged particle swarms in a weakly ionized gas can be suitably modelled by the linear Boltzmann equation. In this work we discuss the time-dependent problem, the stationary problem as well as the long time behaviour of the particle distribution.

The paper is divided in three main parts. The first part is devoted to a simplified one-dimensional Boltzmann model of the Kač type, to study the velocity distribution of a spatially uniform diluted guest population of electrons moving within a host medium under the influence of a D.C. electric field. Necessary conditions and sufficient conditions are established for the existence, uniqueness and attractivity of a stationary nonnegative distribution corresponding to a specified concentration level. Conditions for the onset of the runaway process are established and the long time behaviour of the velocity distribution is studied within the framework of scattering theory.

The second part is devoted to the study of a non-homogeneous model where the collision frequency and the scattering kernel depend also on the space coordinates. A definition of "runaway" and a necessary condition for the suppression of runaways are given. The time-dependent problem is discussed and the long time behaviour of the solution is investigated. Also in this case, under physically reasonable assumptions on the collision frequency, we prove the existence of wave operators and the corresponding existence of travelling waves.

Finally, in the third part, we report some results about three-dimensional velocity systems.

1. INTRODUCTION

In plasma physics, the term "swarm" means a large number of charged particles moving in a background neutral gas. If the particle swarm is characterized by the distribution function $f(\vec{x},\vec{v},t)$ at time t, with $\vec{x}$ the position and $\vec{v}$ the velocity, the basic description of the swarm can be given in terms of the linear Boltzmann equation. Within this context, we consider swarms with only one species of charged particles, electrons or ions, in a single component neutral gas. In the physics of swarms, the collisions of the charged particles with the neutral gas molecules are predominant compared to the collisions between charged particles. In the above case, the plasma gas is said to be weakly ionized and the kinetic theory is based on the following linear Boltzmann equation

$$\frac{\partial f}{\partial t}(\vec{x},\vec{v},t)+\vec{v}\cdot\frac{\partial f}{\partial \vec{x}}(\vec{x},\vec{v},t)+\vec{a}\cdot\frac{\partial f}{\partial \vec{v}}(\vec{x},\vec{v},t)+\nu(\vec{x},|\vec{v}|)f(\vec{x},\vec{v},t)=$$

$$= \int_{\mathbf{R}^3} k(\vec{x},\vec{v},\vec{v}')\nu(\vec{x},|\vec{v}'|)f(\vec{x},\vec{v}',t)d\vec{v}'. \quad (1.1)$$

Here, $\vec{a}$ is the electrostatic field; $\nu(\vec{x},|\vec{v}|)$ is the collision frequency and $k(\vec{x},\vec{v},\vec{v}')$ is the scattering kernel.

Many books have been devoted to swarm physics, but we limit ourselves to quoting only the recent reports by Kumar et al. [1], [2]. These reports can be useful to the researcher interested in the mathematical physics of the problem or to the mathematically oriented kinetic theorist. Problems of this type, other than in the physics of swarms in a background neutral gas, are met in the study of runaway electrons in fully ionized plasmas [3], [4], [5], in the calculation of D.C. conductivity in biological membranes [6], and in the field of electron transport in semiconductors [7], [8].

Much of the literature in this field is devoted to kinetic equations, in differential form, see for instance [9] and [10] and the literature there quoted; however, recently the Boltzmann integro-differential equation has received increasing attention. The purpose of this paper is to study some mathematical aspects of the behaviour of a population of charged particles under the influence of an electric field, using the Boltzmann formulation.

The paper is divided in three main parts. In the first part, after the problem is posed, we study the one-dimensional

Boltzmann model for a spatially uniform diluted guest population of electrons, which corresponds to the Kač model in nonlinear kinetic theory. Two different physical situation are investigated: the "relaxation" case and the "runaway" case. The first case appears when the collision process is sufficiently effective to average kinetic energy and drift velocity, so that particle distribution relaxes to equilibrium. In the second case, the collision process is not effective in removing kinetic energy from the particles, so that no relaxation towards a nonzero steady state profile occurs. In this case, the swarm exhibits a phenomenon called "runaway". "Runaway" is characterized by particles that begin to accelerate or runaway. The key ingredient in distinguishing between these physical situations is the dependence of the collision frequency upon the speed v for large values of v. In fact, if $\nu(v)$ goes to zero too fast as $t\to\infty$, then kinetic energy is not removed from the particles, and these particles run away. In Sections 3 and 4 we assume

$$\int_{-\infty}^{+\infty} \nu(v)dv = +\infty, \tag{1.2}$$

and we study existence, uniqueness and attractivity of a stationary solution. When assumption (1.2) is not satisfied, in Section 5, the existence of wave operators is proved within the framework of scattering theory. Then the long term behaviour of the velocity distribution and the occurrence of travelling waves is studied. This part is principally based on results and proofs found in [11] and [12].

The second part is devoted to the study of a non-homogeneous model where the collision frequency and the scattering kernel depend also on the space coordinates. A definition of "runaway" and a necessary condition for the suppression of runaways are given in Sections 7 and 8. The time-dependent problem is discussed and the long time behaviour of the solution is investigated. Also in this case, under physically reasonable assumptions on the collision frequency, we prove the existence of wave and scattering operators and the corresponding existence of travelling waves, see Sections 9 and 10.

In the third part, we consider a three-dimensional velocity model for a spatially uniform population of electrons. In Sections 11 and 12, a mathematical analysis of the necessary condition for equilibrium is carefully outlined; then, we report some results obtained recently by L. Arlotti on the time-dependent and the stationary problems, [13]. Finally, in Section

13, the scattering theory in the three-dimensional case is presented.

PART I

2. STATEMENT OF THE PROBLEM: THE UNIFORM CASE

Let us consider the one-dimensional model obtained from Eq. (1.1) in a uniform medium, taking account only of the parallel component of the velocity $\vec{v}$. In this approximation, the linear Boltzmann equation for electron swarms in a uniform weakly ionized host medium has the form

$$\frac{\partial f}{\partial t}(v,t) + a\frac{\partial f}{\partial v}(v,t) + \nu(v)f(v,t) = \int_{-\infty}^{+\infty} k(v,v')\nu(v')f(v',t)dv', \quad (2.1)$$

where $f(v,t)$ is the electron distribution as a function of the speed $v \in (-\infty,+\infty)$ and time $t \geq 0$; a is the constant electrostatic acceleration; $\nu(v)$ denotes the collision frequency (between an electron and the host medium) and $k(v,v')$ is the collision kernel, with $k(v,v')dv$ the probability that an electron entering the collision with velocity v' will come out of the collision with its velocity in the interval $(v,v+dv)$. $k(v,v')$ is a nonnegative function, such that

$$\int_{-\infty}^{+\infty} k(v,v')dv \equiv 1, \qquad \forall v' \in \mathbf{R}.$$

In this model, the electron-electron interaction and ion-electron interaction are neglected and the processes of recombination and ionization are assumed to balance each other. Equation (2.1) must be supplemented with an initial condition of the form

$$f(v,0) = f_0(v). \quad (2.2)$$

Because the physics of the problem requires a finite electron concentration and a finite collision rate (per unit volume), it is appropriate to introduce the Banach spaces $L_1(\mathbf{R},dv)$ and $L_1(\mathbf{R},\nu dv)$, with the norms

$$\|f\|_1 = \int_{-\infty}^{+\infty} |f(v)|dv,$$

$$\|f\|_\nu = \int_{-\infty}^{+\infty} \nu(v)|f(v)|dv.$$

Before writing equations (2.1)-(2.2) as an abstract initial-value problem, we make the following assumptions on a, $\nu(v)$ and $k(v,v')$.

Assumption 1. The acceleration a is a fixed positive constant.

Assumption 2. The collision frequency $\nu(v)$ is a Lebesgue measurable, nonnegative and even function of v on $(-\infty,+\infty)$, which is almost everywhere nonzero and Lebesgue integrable on every bounded Lebesgue measurable set.

Assumption 3. The collision kernel satisfies the reciprocity symmetry property

$$k(-v,-v') = k(v,v').$$

The operator K formally defined by

$$(Kf)(v) = \int_{-\infty}^{+\infty} k(v,v')\nu(v')f(v')dv' \tag{2.3}$$

is a positive linear operator $K: L_1(\mathbf{R},\nu dv)\to L_1(\mathbf{R},dv)$ satisfying

$$\|Kf\|_1 = \|f\|_\nu, \quad \text{if } f \in L_1(\mathbf{R},\nu dv) \text{ and } f \geq 0.$$

Moreover, let us define the following operators:

$$\begin{cases} T_0: D(T) \subset L_1(\mathbf{R},dv) \to L_1(\mathbf{R},dv) \\ \qquad T_0 f = -a\dfrac{\partial f}{\partial v}, \end{cases}$$

$$\begin{cases} A: L_1(\mathbf{R},\nu dv) \cap L_1(\mathbf{R},dv) \to L_1(\mathbf{R},dv) \\ \qquad Af = -\nu(v)f, \end{cases}$$

$$\begin{cases} T: D(T) \subset L_1(\mathbf{R},dv) \to L_1(\mathbf{R},dv) \\ \qquad Tf = T_0 f + Af + Kf \end{cases}$$

where $D(T) = L_1(\mathbf{R},dv) \cap L_1(\mathbf{R},\nu dv) \cap \mathfrak{D}$ and $\mathfrak{D}$ is the set of functions which are absolutely continuous on [-b,b] for all b>0, are of bounded variation and vanish at $-\infty$.

Now, using the preceding definitions, we can rewrite equations (2.1)-(2.2) into the more concise form

$$\frac{df}{dt}(t) = Tf(t), \qquad t>0, \tag{2.4}$$

$$\lim_{t\to 0} f(t) = f_0, \tag{2.5}$$

where d/dt is a strong derivative, $f(t) = f(\cdot,t)$ is a function from $[0,+\infty)$ into $L_1(\mathbf{R},dv)$ and f_0 is the initial datum. The operator T defined on D(T) is generally not a closed operator. This property depends strongly upon the form of the collision operator K, but we can prove that an extension of T is the generator of a strongly continuous semigroup on $L_1(\mathbf{R},dv)$. This result does not rely directly on the abstract theory of time dependent kinetic equations of Beals and Protopopescu (see Ref.

[14]; also Ref. [15], Ch. XI and Sections XII.1-2), because, in general, we have an unbounded collision frequency and an unbounded gain part of the collision operator. A direct proof based on the Hille-Phillips theorem has been done in [11]; the result is summarized in the following lemma, (see [11], Theorem 7).

LEMMA 1. For every $\lambda>0$ and every arbitrary function $g\in L_1(\mathbf{R},dv)$ there exists a unique solution $T_\lambda g$ of the resolvent equation $(\lambda-T)f=g$, which belongs to $L_1(\mathbf{R},dv)$. Then T_λ is the resolvent of a strongly continuous positive contraction semigroup $\{S(t);\ t\geq 0\}$ on $L_1(\mathbf{R},dv)$, whose generator G is a closed extension of T. Moreover $\{S(t);\ t\geq 0\}$ satisfies

$$\|S(t)f\|_1 = \|f\|_1, \qquad f \geq 0 \tag{2.6}$$

if and only if G is the (minimal) closure of T.

In many cases, the generator G is the minimal closure of T, for instance, if $\nu(v)$ is essentially bounded or if $\nu(v)$ is integrable. Another case when (2.6) is true occurs if there exists a nonzero stationary solution (see next section). From now on, we will not distinguish between the operators T and T_0 and their closed extensions (or closures) in $L_1(\mathbf{R},dv)$. Let us denote by $W_0(t)$ the evolution group generated by the free streaming operator T_0 given by

$$[W_0(t)g](v) = g(v-at), \qquad t\in\mathbf{R},$$

and by $S_0(t)$ the semigroup

$$[S_0(t)g](v) = \exp\Big(-\int_0^t \nu(v-as)\,ds\Big)g(v-at),$$

generated by the streaming operator $T_0-\nu$.

3. THE STATIONARY PROBLEM

In the Introduction, we discussed briefly the importance of the behaviour of $\nu(v)$ for large v, in order to characterize the long-time behaviour of the solution to problem (2.1)-(2.2). Moreover, the non integrability of $\nu(v)$ over $\mathbf{R}$ is a well-known necessary condition for the existence of the steady-state distribution, [16], [17], [18]. This condition corresponds to requiring that the probability of a time of flight exceeding time t, when the electron starts freely with speed v, tends to zero as $t\to\infty$. In this case the collision frequency cannot vanish too fast, when $v\to\infty$, [17], [19]. For this reason, in order to study the stationary problem in this section, we make the additional assumption

$$\int_{-\infty}^{+\infty} \nu(v)dv = +\infty. \tag{3.1}$$

Eq. (3.1) characterizes the frequency behaviour of $\nu(v)$ as $v\to\infty$, because the function $\nu(v)$ is locally integrable. Thus, (3.1) is equivalent to

$$\int_{\alpha}^{+\infty} \nu(v)dv = +\infty, \qquad \text{for all } \alpha \in \mathbf{R}.$$

Let us consider now the time-independent problem

$$0 = T_0f + Af + Kf. \tag{3.2}$$

By a solution of the time-independent equation (3.2), we mean a function $\varphi=\varphi(v)$ satisfying (3.2) and belonging to $L_1(\mathbf{R},\nu dv)$. This solution will be called a stationary one. The derivative appearing in (3.2) is taken in the distributional sense, so that each stationary solution will be absolutely continuous on every bounded set in $\mathbf{R}$. Recalling the physical meaning of the function φ, we require that the stationary solution is nonnegative and belongs to $L_1(\mathbf{R},dv)$. It will soon appear that every φ satisfies $\varphi(-\infty)= \varphi(+\infty)=0$. The integrodifferential equation (3.2) can be easily transformed into an (equivalent) integral one. In order to attain the above, we define the following operator

$$\begin{cases} L: L_1(\mathbf{R},dv) \to L_1(\mathbf{R},\nu dv) \\ (Lf)(v) = \int_{-\infty}^{v} \frac{1}{a} \exp\{-\frac{1}{a}\int_{v'}^{v} \nu(v'')dv''\}f(v')dv'. \end{cases}$$

Under assumption (3.1), L is a positive contraction from $L_1(\mathbf{R},dv)$ to $L_1(\mathbf{R},\nu dv)$, LK is also a positive contraction on $L_1(\mathbf{R},\nu dv)$, and $\|Lf\|_\nu = \|f\|_1$ for all nonnegative $f \in L_1(\mathbf{R},dv)$. Thus we have

LEMMA 2. Under assumption (3.1) of non-integrability of $\nu(v)$, every solution of integrodifferential equation (3.2) in $L_1(\mathbf{R},\nu dv)$ is a solution of the linear integral problem

$$\varphi = LK\varphi, \quad \varphi \in L_1(\mathbf{R},\nu dv), \tag{3.3}$$

and viceversa. Therefore, for every solution φ of the two equivalent problems we have $\varphi(-\infty)=\varphi(+\infty)=0$.

<u>Proof.</u> See [11], Theorem 2. We may limit ourselves to $\varphi(\pm\infty)=0$ due to the fact that φ is of bounded variation on $\mathbf{R}$ in

combination with assumption (3.1). □

For the sake of convenience, we report the explicit form of (3.3)

$$\varphi(v) = \frac{1}{a}\int_{-\infty}^{v} \exp\{-\frac{1}{a}\int_{v'}^{v} \nu(v'')dv''\} \int_{-\infty}^{+\infty} k(v',\hat{v})\nu(\hat{v})\varphi(\hat{v})d\hat{v}\, dv', \qquad v\in\mathbf{R}.$$

It is worth noting that the integral equation equivalent to the integrodifferential equation takes the form (3.3) only if condition (3.1) is satisfied. In fact, if the alternative assumption

$$\int_{-\infty}^{+\infty} \nu(v)dv < +\infty$$

is satisfied, then the integrodifferential equation (3.2) takes the new form

$$\varphi(v) = (LK\varphi)(v) + \exp\{-\frac{1}{a}\int_{-\infty}^{v} \nu(v')dv'\}\varphi(-\infty). \tag{3.4}$$

In this case, we obtain the existence of the continuous limits $\varphi(\pm\infty)$ which turn out to be finite, and $\varphi(-\infty)=\varphi(+\infty)$. The "new" stationary problem (3.4) has a unique nonnegative solution φ in $L_1(\mathbf{R},\nu dv)$ of unit norm with $\varphi(-\infty)=\varphi(+\infty)>0$, but this solution is physically irrelevant, since it does not belong to $L_1(\mathbf{R},dv)$, which corresponds to an infinite population level.

Up to now, we did not make assumptions on the collision operator. Generally, for different physical situations the kernels appearing in the collision operator are determined from experiments and they are very complicated functions of velocities. Without putting conditions on the operator K, we are not able to give an existence theorem for the problem (3.3). We can only prove that the set of all φ satisfying problem (3.3) is at most one dimensional and, when nontrivial, contains a nontrivial nonnegative function, ([11], Theorem 3). The most general result on the existence of nonnegative solutions for the integral problem (3.3) is given in the following theorem, [11].

THEOREM 3. If assumption (3.1) is satisfied and if, in addition, LK is a weakly compact operator on $L_1(\mathbf{R},\nu dv)$, then the stationary integral problem (3.3) has a unique nonnegative solution in $L_1(\mathbf{R},\nu dv)$ of unit norm.

<u>Proof</u>: This result is a consequence of the Krasnosel'skiĭ theory for the existence of a positive eigenvector for positive opera-

tors, (see [20], Chapter 2). The operator LK has the spectral radius, spr(LK), equal to one, because, under assumption (3.1) and f nonnegative, we have $\|LKf\|_\nu = \|f\|_\nu$. Moreover, $(LK)^2$ is compact as an operator on $L_1(\mathbf{R},\nu dv)$, because the square of a weakly compact operator in L_1 is compact. Thus the compactness of $(LK)^2$ in combination with spr(LK)=1 ensures the existence of at least one nonnegative $\varphi \in L_1(\mathbf{R},\nu dv)$ of unit norm. The uniqueness follows directly from the previous lemma. □

As a corollary, we could require that the operator

$$(Bf)(v) = \int_{-\infty}^{+\infty} k(v,v')f(v')dv'$$

is weakly compact on $L_1(\mathbf{R},dv)$, because ν is a bounded operator from $L_1(\mathbf{R},\nu dv)$ into $L_1(\mathbf{R},dv)$. Finally, the previous theorem is trivially satisfied whenever K, defined by (2.3), is a compact operator and this is the case of many practical situations. For example, the collision operator in the BGK model is compact as an operator from $L_1(\mathbf{R},\nu dv)$ to $L_1(\mathbf{R},dv)$, and, more generally, each collision operator with a separable kernel is compact. Such operators have a kernel which is a finite linear combination of essentially bounded functions separated in the variables v and v′. Besides producing substantial semplifications in the mathematical treatment of this problem, these types of kernels fit experimental values with sufficient accuracy, [21], [22].

4. DECAY TO EQUILIBRIUM

In this section, we will investigate the long-time behaviour of the solution when the stationary solution exists. Therefore, here we make the assumption (3.1) and we will examine the other case, when (3.1) is not true, in the next section.

First, we will report some results on the time-dependent problem. See [11] for the proofs.

THEOREM 4. Let us suppose a nontrivial solution φ of the stationary problem in $L_1(\mathbf{R},dv)$ exists. Then the semigroup S(t) generated by (the closure of) T is mean ergodic, i.e., for every $g \in L_1(\mathbf{R},dv)$, there exists a vector $Pg \in L_1(\mathbf{R},dv)$ such that

$$\lim_{t\to\infty} \left\| \frac{1}{t} \int_0^t S(t')g\,dt' - Pg \right\|_1 = 0;$$

the limit Pg is a one dimensional projection of the form

$$(Pg)(v) = \alpha(g)\varphi(v), \quad v\in\mathbf{R},$$

where

$$\alpha(g) = \int_{-\infty}^{\infty} \psi(v')g(v')\,dv'$$

for some function $\psi\in L_\infty(\mathbf{R},dv)$ with $\psi\geq 0$, $\int_{-\infty}^{+\infty} \psi(v')\varphi(v')\,dv'=1$ and $\|\psi\|_\infty<+\infty$.

An important necessary requirement for having the convergence of the time dependent solution to a stationary solution is that the generator T of the semigroup S(t) does not have purely imaginary eigenvalues. Without entering into details, this condition is supported by the fact that the solution should not contain a term $\exp(i\alpha t)f_0$, where $i\alpha$ is the purely imaginary eigenvalue and f_0 is the initial datum, since such a factor does not converge as $t\to\infty$. The absence of such eigenvalues implies $\ker T=\bigcap_{t\geq 0}\ker(I-S(t))$, (see [23], A-III, Cor. 6.4). This fact assures the convergence of S(t)g on $\ker(I-S(t))$. In the present case, when the generator G is the minimal closure of T and it is again denoted by T, the following lemma can be proved, [13].

LEMMA 5. The set of purely imaginary eigenvalues of the generator T of S(t) is empty.

<u>Proof.</u> Let $i\lambda$ be a purely imaginary eigenvalue of T with corresponding eigenvector f. Then we have $S(t)f = e^{i\lambda t}f$ for all $t\geq 0$ and $S(t)|f| = |f|$ for all $t\geq 0$, because the L_1-norm is strictly monotone on the positive cone L_1^+ (cf. [23], Cor. 2.3 at p. 297). Using the Duhamel formula,

$$|f| = S(t)|f| = S_0(t)|f| + \int_0^t S_0(s)K|f|\,ds,$$

as well as

$$f = e^{-i\lambda t}S_0(t)f + \int_0^t e^{-i\lambda s}S_0(s)Kf\,ds,$$

is obtained. Comparing these equations we find $|Kf|=K|f|$. Then, following the same procedure of [13], we obtain $\lambda=0$. □

Now we give the main theorem on the asymptotic behaviour of the evolution operator S(t), when a stationary solution exists.

THEOREM 6. Let us suppose that a nontrivial stationary solution exists. Then for every $g\in L_1(\mathbf{R},dv)$

$$\lim_{t\to\infty} \|S(t)g - Pg\|_1 = 0.$$

A detailed proof of this theorem can be found in [11]. Due to lemma 5, the hypothesis of no purely imaginary eigenvalues can be dropped. We observe that generally $\lim_{t\to\infty} S(t)$ is always a projection P onto the fixed space of S(t) which coincides with the kernel of T (when P=0 there is stability, but this is not the case). When a nontrivial positive fixed function exists, which belongs to ker T (in this case, a stationary solution), the so-called "0-2 law" can be used successfully to study the convergence of positive semigroups. The proof of Theorem 6 is essentially based on the "0-2 law" for positive operators, ([23], C-IV Th. 2.6 and Cor. 2.7).

5. TRAVELLING WAVES IN THE ELECTRON TRANSPORT PROBLEM

In this section, we will examine the asymptotics of the electron transport problem if assumption (3.1) is not satisfied. When

$$\int_{-\infty}^{+\infty} \nu(v)\,dv < +\infty, \tag{5.1}$$

the stationary solution does not exist in $L_1(\mathbf{R},dv)$, but an unphysical solution will exist in $L_1(\mathbf{R},\nu dv)$. In this case, no relaxation to equilibrium occurs, and the charged particles are accelerated indefinitely. Here the so-called runaway phenomenon appears and the collisions generate a travelling wave in velocity space with "velocity" a.

Hoping to prove that the solution $[S(t)g](v)$ behaves like a wave for large t, i.e. an operator Ω^- (independent of t) such that $[S(t)g](v) \simeq [\Omega^- g](v-at)$ as t goes to infinity exists. Scattering theory, applied to the electron transport problem, seems to be the suitable tool for proving the existence of travelling waves in velocity space. Therefore, we prove the existence of the so-called wave operators under the assumption of integrability of $\nu(v)$ on $\mathbf{R}$.

Scattering theory is used to study the large time behaviour of many dynamical systems and it was applied to neutron transport theory by Hejtmanek and Simon ([24], [25]; see also [26], [27]). Here we report some results of [12], where this theory is used in electron transport theory to exhibit travelling waves in the runaway regime. Scattering theory compares the free dynamics (in the present problem, generated by the group $W_0(t)$) and the full dynamics (generated by the semigroup $S(t)$). Such comparison is performed by the wave operators, defined by the following limits in the strong operator topology of $L_1(\mathbf{R},dv)$:

$$\Omega^+ = \underset{t\to-\infty}{\text{s-lim}}\; S(-t)W_0(t),$$

$$\Omega^- = \underset{t\to+\infty}{\text{s-lim}}\; W_0(-t)S(t).$$

The existence of wave operators allows a comparison between the free solution $W_0(t)g$ corresponding to the initial datum g and the full motion $S(t)h$ for suitable initial datum h as $t\to\infty$, thereby finding the mathematical link between g and h. We note that these definitions involve $S(t)$ only for positive times; they differ from the usual ones, because $S(t)$ is generally not a group.

In this section, the following theorem on the existence of the wave operators Ω^- and Ω^+ is presented without a proof, because it will be dealt with later in Section 9. Emphasis should be placed on the important role played by the integrability of the collision frequency $\nu(v)$ on $\mathbf{R}$, which sufficiently ensures the existence of Ω^-.

THEOREM 7. If the collision frequency $\nu(v)$ satisfies Assumption 2 and the additional assumption (5.1), then the limits

$$\Omega^- = \underset{t\to+\infty}{\text{s-lim}}\; W_0(-t)S(t),$$

$$\Omega^+ = \underset{t\to-\infty}{\text{s-lim}}\; S(-t)W_0(t)$$

exist in the strong operator topology of $L_1(\mathbf{R},dv)$ and are bounded positive operators.

Proof. The proof will be given in its more general context (non uniform systems) in Section 9. □

In the remaining part of this section, we will prove that the solution of the time dependent problem will increasingly resemble a travelling wave in velocity space with "velocity" a as $t\to+\infty$.

A travelling wave (in velocity space) of "velocity" a means a function of the form $W_0(t)g$, where g does not depend on t, i.e. a function of the form

$$[W_0(t)g](v) = g(v-at).$$

The main result can be summarized as follows:

THEOREM 8. Suppose that $\nu(v)$ satisfies Assumption 2 and condition (5.1). Then for every $g \in L_1(\mathbb{R},dv)$

$$\lim_{t\to+\infty} \|W_0(-t)S(t)g - \Omega^- g\|_1 = 0,$$

where Ω^- is the wave operator defined above.

Proof. See [12], where asymptotic properties of the streaming operator $S_0(t)$ are given. □

In order to accomplish the objective, take note that from the previous theorem

$$[S(t)g](v) \simeq [\Omega^- g](v-at), \qquad \text{as} \quad t\to+\infty,$$

i.e., physically speaking, in the remote future, $S(t)g$ looks like a travelling wave in velocity space.

PART II

6. THE NON UNIFORM CASE

The second part of this paper will be devoted to the study of the one dimensional model in the non uniform case, taking into account the component of $\vec{x}$ along the direction of the acceleration $\vec{a}$ and only the parallel component of the velocity $\vec{v}$. The space-velocity electron distribution $f(x,v,t)$ in a weakly ionized host medium obeys the linear integro-differential equation

$$\frac{\partial f}{\partial t}(x,v,t) + v\,\frac{\partial f}{\partial x}(x,v,t) + a\,\frac{\partial f}{\partial v}(x,v,t) + \nu(x,v)f(x,v,t) =$$

$$= \int_{-\infty}^{+\infty} k(x,v,v')\nu(x,v')f(x,v',t)dv'. \qquad (6.1)$$

In this model the electrostatic acceleration is assumed to be uniform in time and position, i.e. a is constant and positive.

Recombination and ionization effects are assumed to balance each other. The collision frequency $\nu(x,v)$ and the scattering kernel $k(x,v,v')$ do not depend on the temporal variable. The evolution equation (6.1) is equipped with the initial condition

$$f(x,v,0) = f_0(x,v). \tag{6.2}$$

Let us introduce $L_1(\mathbf{R}^2,dxdv)$ as the functional setting where the problem (6.1)-(6.2) is studied, with the norm

$$\|f\|_1 = \int_{-\infty}^{+\infty}\int_{-\infty}^{+\infty} |f(x,v)|\, dxdv.$$

Let us also introduce the Banach space $L_1(\mathbf{R}^2,\nu(x,v)dxdv)$, with the norm

$$\|f\|_\nu = \int_{-\infty}^{+\infty}\int_{-\infty}^{+\infty} \nu(x,v)|f(x,v)|\, dxdv.$$

Referring to the assumptions made for the uniform problem, Assumption 1 still remains valid, whereas the new assumptions follow:

Assumption 2'. There exists a function $\bar\nu=\bar\nu(v)$ such that the collision frequency ν satisfies

$$0 \le \nu(x,v) \le \bar\nu(v), \qquad \text{for all } x\in\mathbf{R},$$

where $\bar\nu(v)$ is a Lebesgue measurable, nonnegative and even function of v on $(-\infty,+\infty)$, which is almost everywhere nonzero and Lebesgue integrable on every bounded Lebesgue measurable set.

Assumption 3'. The collision kernel $k(x,v,v') \ge 0$, appearing in the integral operator, is such that

$$\int_{-\infty}^{+\infty} k(x,v,v')\, dv = 1, \qquad \text{a.e. } x\in\mathbf{R}, \qquad \forall v'\in\mathbf{R},$$

and by reciprocity symmetry, we also have $k(x,-v,-v')=k(x,v,v')$, a.e. $x\in\mathbf{R}$.

Noting that Assumption 2' is not restrictive, because the total collision frequency depends mainly on the speed; here singularities can arise only with respect to the variable v. When recombination and ionization balance each other, the solution to (6.1) has to be

$$n_0(x,t) = \int_{-\infty}^{+\infty} f(x,v,t)\, dv = \int_{-\infty}^{+\infty} f_0(x,v)\, dv,$$

where $n_0(x,t)$ is the number density.

7. NOTATIONS, DEFINITIONS, AND PRELIMINARY REMARKS

In this section we want to define some of the principal physical phenomena arising in the electron transport modelled by (6.1) and (6.2)

In the first part of this paper, we discussed the runaway phenomenon that arises in the evolution of electron swarms. This appears when some particles increase their speed indefinitely, until they run away. Essentially, this phenomenon occurs in velocity space. Thus, as a simple criterion for the suppression of runaway, we used the non integrability of $\nu(v)$. This criterion is commonly found in the literature of plasma physics. A (not heuristic) proof was given by Cavalleri and Paveri-Fontana, [18], assuming the existence of a steady state. Another similar proof, working on the integral form of the Boltzmann equation, is found in [11]. Here, it is necessary to revise the criterion in light of the more general context. First we characterize the runaway phenomenon in a system with spatial dependence and we give the following definitions.

Let us introduce the velocity electron distribution,

$$\mathcal{F}(v,t)= \int_{-\infty}^{+\infty}\varphi(x,v,t)\; dx, \tag{7.1}$$

where $\varphi=\varphi(x,v,t)$ is the solution of problem (6.1). Here a unique and sufficiently smooth solution to the problem is supposed to exist.

DEFINITION 1. The physical process modelled by equation (6.1) gives rise to runaway, if there exists an integrable function $\mathcal{F}_\infty(v)$ such that

$$\lim_{t\to+\infty} \mathcal{F}(v+at,t) = \mathcal{F}_\infty(v).$$

Here $\mathcal{F}(v,t)$ is defined by (7.1). The limit is taken in the sense of the norm of $L_1(\mathbf{R},dv)$ and, at present, no further properties on $\mathcal{F}_\infty$ are required. A different, more hydrodynamic, characterization of runaway, can be also given by making use of the concept of average speed

$$\langle v\rangle(t) = \frac{\displaystyle\int_{-\infty}^{+\infty}dx\int_{-\infty}^{+\infty}dv\; v\;\varphi(x,v,t)}{\displaystyle\int_{-\infty}^{+\infty}dx\int_{-\infty}^{+\infty}dv\;\varphi(x,v,t)}.$$

DEFINITION 2. The physical process modelled by equation (6.1) gives rise to runaway, if

$$\lim_{t\to+\infty} \langle v\rangle(t) = \infty.$$

If we define

$$\langle v\rangle_\infty = \lim_{t\to+\infty} \langle v\rangle(t),$$

in the general case, $\langle v\rangle_\infty$ is a positive real number. When $\langle v\rangle_\infty \neq \infty$, the physical process modelled by equation (6.1) does not give rise to runaway, and we cannot say if the process decays to equilibrium. Definitions 1 and 2 are not necessarily equivalent. In fact, runaway can be also defined in other ways, but this is not the objective of the present paper.

Before giving an abstract formulation to problem (6.1)-(6.2), we denote by T_0, A and K the following operators:

$$\begin{cases} T_0 : D(T_0\ ) \subset L_1(\mathbf{R}^2, dxdv) \to L_1(\mathbf{R}^2, dxdv) \\ T_0 f = -\, v\,\dfrac{\partial f}{\partial x} - a\,\dfrac{\partial f}{\partial v}\ , \end{cases}$$

$$\begin{cases} A:\ L_1(\mathbf{R}^2, \nu dxdv) \cap L_1(\mathbf{R}^2, dxdv) \to L_1(\mathbf{R}^2, dxdv) \\ Af = -\,\nu(x,v) f, \end{cases}$$

$$\begin{cases} K:\ L_1(\mathbf{R}^2, \nu dxdv) \to L_1(\mathbf{R}^2, dxdv) \\ (Kf)(v) = \displaystyle\int_{-\infty}^{+\infty} k(x,v,v')\nu(x,v')f(x,v')dv'. \end{cases}$$

K is a positive linear operator satisfying

$$\|Kf\|_1 = \|f\|_\nu, \qquad \text{if } f \in L_1(\mathbf{R}^2, \nu dxdv) \text{ and } f \geq 0.$$

Using the preceding definitions, problem (6.1)-(6.2) can be put into the following abstract form

$$\frac{df}{dt}(t) = T_0 f\ (t) + Af\ (t) + Kf\ (t)\ , \qquad t > 0, \tag{7.2}$$

$$\lim_{t\to 0} f(t) = f_0, \tag{7.3}$$

formally similar to (2.4)-(2.5), where $f(t) = f(\cdot,t)$ is a function from $[0,+\infty)$ into $L_1(\mathbf{R}^2, dxdv)$ and f_0 is the initial datum.

The existence theory for problem (7.2)-(7.3) does not fall directly within the general abstract kinetic theory of [15] and [14], because the collision frequency (and thus the scattering operator) is unbounded. We are not interested in this generalization, which can be performed working in the functional L_1-setting, following [28]. In this paper we wish to limit ourselves to the scattering theory for the electron problem and to the study of the travelling waves, and hence we prefer to leave the existence theory out of consideration. In the remaining part of this section, notations are given and some results are assumed without a detailed proof.

First of all, the characteristic lines corresponding to the collisionless problem, are introduced by

$$\begin{cases} \xi = x - vt + \frac{1}{2}at^2, \\ \eta = v - at. \end{cases}$$

Now the free semigroup $W_0(t)$: $L_1(\mathbf{R}^2, dxdv) \to L_1(\mathbf{R}^2, dxdv)$ can be defined by

$$(W_0(t)\ g)\ (x,v) = g\left(x-vt+\tfrac{1}{2}at^2,\ v-at\right)$$

and consequently the semigroup generated by (the closure of) T_0+A

$$S_0(t)g(x,v) = \exp\left(-\int_0^t \nu(x-vs+\tfrac{1}{2}as^2, v-as)ds\right)\ g\left(x-vt+\tfrac{1}{2}at^2, v-at\right).$$

We denote by $S(t)$ the full semigroup generated by (a closed extension of) the operator $T=T_0+A+K$, with domain $D(T)= D(T_0) \cap L_1(\mathbf{R}^2, \nu dxdv)$ and $D(T_0)=\{f: f\in L_1(\mathbf{R}^2, dxdv),\ \ T_0 f \in L_1(\mathbf{R}^2, dxdv)\}$, where $T_0 f$ is defined in the distributional sense. Following the above, $S(t)$ can be assumed as a perturbation of $S_0(t)$, satisfying

$$\|S(t)g\|_1 = \|g\|_1, \quad \text{for nonnegative } g, \tag{7.4}$$

and, when $\bar{\nu}(v)$ is integrable, the two Duhamel formulas can be derived.

8. ON THE NECESSARY CONDITION FOR EQUILIBRIUM

Now we give a proof of a necessary condition for the existence of the steady-state solution to problem (6.1)-(6.2), when the collision frequency satisfies Assumption $2'$. Even if this proof is formally different from the previous ones given in [18] and [11], it is also based on the mathematical requirement (with obvious physical meaning) that the solution must be an L_1-integrable function.

THEOREM 9. Let us suppose that a, ν and K satisfy the Assumptions 1, $2'$, and $3'$. A necessary condition for the existence of a nontrivial nonnegative stationary solution to problem (6.1)-(6.2) in $L_1(\mathbf{R}^2, dxdv) \cap L_1(\mathbf{R}^2, \nu dxdv)$ is that

$$\int_{-\infty}^{+\infty} \bar{\nu}(v)dv = +\infty, \tag{8.1}$$

for any $\bar{\nu}$ which satisfies Assumption $2'$.

Proof. Let us suppose that the time-independent problem corresponding to (6.1) admits a nontrivial and sufficiently smooth nonnegative solution $\varphi \in L_1(\mathbf{R}^2, dxdv) \cap L_1(\mathbf{R}^2, \nu dxdv)$. Then the solution satisfies the following integral equation

$$\varphi(x,v) = \int_0^{+\infty} \exp\Big(-\int_0^t \nu(x-vs+\tfrac{1}{2}as^2, v-as)ds\Big)\cdot$$

$$\cdot\int_{-\infty}^{+\infty} k(x-vt+\tfrac{1}{2}at^2, v-at, \hat{v})\nu(x-vt+\tfrac{1}{2}at^2, \hat{v})\ \varphi(x-vt+\tfrac{1}{2}at^2, \hat{v})d\hat{v}dt.$$

Suppose for the moment that for some $\bar{\nu}$ which satisfies Assumption $2'$

$$\int_{-\infty}^{+\infty} \bar{\nu}(v)dv = M < +\infty, \tag{8.2}$$

we will prove that this is an absurdum. Manipulating the above equation we also have

$$\varphi(x,v) \geq \exp\Big(-\frac{1}{a}\int_{-\infty}^{+\infty}\bar{\nu}(w)dw\Big)\cdot$$

$$\cdot\int_0^{+\infty}\int_{-\infty}^{+\infty} k(x-vt+\tfrac{1}{2}at^2, v-at, \hat{v})\nu(x-vt+\tfrac{1}{2}at^2, \hat{v})\ \varphi(x-vt+\tfrac{1}{2}at^2, \hat{v})d\hat{v}dt.$$

Taking the L_1-norm and using (8.2) yield

$$\|\varphi\|_1 \geq \exp\Big(-\frac{M}{a}\Big)\cdot$$

$$\cdot\int_{-\infty}^{+\infty}dx\int_{-\infty}^{+\infty}dv\int_0^{+\infty}dt\int_{-\infty}^{+\infty}k(x-vt+\tfrac{1}{2}at^2, v-at, \hat{v})\nu(x-vt+\tfrac{1}{2}at^2, \hat{v})\varphi(x-vt+\tfrac{1}{2}at^2, \hat{v})d\hat{v}.$$

After making the change of variables $x'=x-vt+\frac{1}{2}at^2$, $v'=v-at$, using Assumption $3'$ and that $\varphi \in L_1(\mathbf{R}^2, \nu dxdv)$, we find that

$$\|\varphi\|_1 \geq \exp\Big(-\frac{M}{a}\Big)\int_{-\infty}^{+\infty}dx'\int_{-\infty}^{+\infty}dv'\int_0^{+\infty}dt\int_{-\infty}^{+\infty}k(x', v', \hat{v})\nu(x', \hat{v})\ \varphi(x', \hat{v})d\hat{v} =$$

$$= \exp\Big(-\frac{M}{a}\Big)\int_0^{+\infty}\|\varphi\|_\nu dt = \infty.$$

We have contradicted the assumption that the solution belongs to L_1, so the integral in (8.2) cannot be finite. Hence the necessary condition (8.1) is proved. □

In the previous proof, we utilized the integral formulation of the stationary problem. The integral theory has been used successfully to study time dependent problems and time independent problems in the field of transport in ionized gases. We only cite the papers by Molinet [29], Boffi et al. [22], and Braglia

[19] and the references quoted therein. However, we think that a more refined analysis of the relationship between the time dependent problem and the stationary one is needed. A rigorous analysis of this connection is not done here and it will be the objective of future work.

9. THE EXISTENCE OF THE WAVE OPERATORS Ω^- AND Ω^+

In this section the existence of the wave operators as defined in Section 5 under Asssumption $2'$ on $\nu(x,v)$ and the sufficient condition of integrability of $\bar{\nu}(v)$ on $\mathbf{R}$ is proved. The simpler case of transport equation in a uniform medium is treated in depth in [12]. Here, using Cook's method [30] generalized to the case of a Banach space, we completely solve the more general case of a non uniform medium.

Assumption $2'$ is very powerful from a mathematical point of view, because it allows us to make many estimates which will be used in the following proofs.

THEOREM 10. If the collision frequency $\nu(x,v)$ satisfies Assumption $2'$ and the additional assumption

$$\int_{-\infty}^{+\infty} \bar{\nu}(v)\,dv = M < +\infty, \tag{9.1}$$

then the limit

$$\Omega^- = \underset{t\to+\infty}{\text{s-lim}}\; W_0(-t)\,S(t) \tag{9.2}$$

exists in the strong operator topology of $L_1(\mathbf{R}^2,dxdv)$ and is a bounded positive operator.

Proof. Assumption $2'$ and (9.1) allow us to easily make the following estimate

$$\int_0^t \nu(x-vs+\tfrac{1}{2}as^2, v-as)\,ds \le \int_0^t \bar{\nu}(v-as)\,ds \le \frac{M}{a} \quad \text{for all} \quad t\ge 0.$$

For all nonnegative $g\in L_1(\mathbf{R}^2,dxdv)$ and for $t\ge 0$, ignoring the contribution of the collision term, we obtain that

$$[S(t)\,g](x,v) \ge \exp\Big(-\int_0^t \bar{\nu}(v-as)\,ds\Big)[W_0(t)\,g](x,v) \;\ge\; \exp\Big(-\frac{M}{a}\Big)[W_0(t)\,g](x,v)$$

and also for a.e. $(x,v)\in\mathbf{R}^2$ and $t\ge 0$

$$g(x,v) \;\le\; [S(t)\,g](x+vt+\tfrac{1}{2}at^2, v+at)\exp\Big(\frac{M}{a}\Big).$$

Replacing g by $S(s)g$ and t by $t-s$ with $t\ge s\ge 0$ gives

$$[S(s)g](x,v) \le [S(t)g](x+v(t-s)+\tfrac{1}{2}a(t-s)^2, v+a(t-s))\exp\left(\tfrac{M}{a}\right).$$

Now this relation is used to give an estimate of the norm of $S(s)g$. Let t be fixed and positive and let g be nonnegative and belonging to $L_1(\mathbf{R}^2, dxdv)$.

$$\int_0^t \|S(s)g\|_\nu ds \le \int_0^t \int_{-\infty}^{+\infty}\int_{-\infty}^{+\infty} \nu(x,v)\ [S(t)g](x+v\tau+\tfrac{1}{2}a\tau^2, v+a\tau)\exp\left(\tfrac{M}{a}\right)dxdv\ d\tau$$

$$\le \exp\left(\tfrac{M}{a}\right)\int_0^t \int_{-\infty}^{+\infty}\int_{-\infty}^{+\infty} \nu(x'-v'\tau+\tfrac{1}{2}a\tau^2,\ v'-a\tau)[S(t)g](x',v')\,dx'dv'd\tau$$

is obtained by changing the variables with $x' = x+v\tau+\frac{1}{2}a\tau^2$, $v'=v+a\tau$. Then using Assumption 2′, (9.1), and the semigroup property (7.4) yields

$$\int_0^{+\infty} \|S(s)g\|_\nu ds \le \exp\left(\tfrac{M}{a}\right)\int_{-\infty}^{+\infty}\int_{-\infty}^{+\infty}\int_0^{+\infty} \bar{\nu}(v'-a\tau)\ d\tau\ [S(t)g](x',v')\,dx'dv' \le$$

$$\le \exp\left(\tfrac{M}{a}\right)\ \|S(t)g\|_1 \int_{-\infty}^{+\infty} \frac{\bar{\nu}(w)}{a}dw = \frac{M}{a}\exp\left(\tfrac{M}{a}\right)\|g\|_1 < +\infty. \qquad (9.3)$$

Now we are in position to prove the existence of Ω^-. First we compute

$$[W_0(-t)S_0(t)g](x,v) = [S_0(t)g](x+vt+\tfrac{1}{2}at^2, v+at) =$$

$$= \exp\left(-\int_0^t \nu(x+v\tau+\tfrac{1}{2}a\tau^2, v+a\tau)d\tau\right) g(x,v).$$

Then the limit in the strong operator topology of $L_1(\mathbf{R}^2, dxdv)$ is easily given by

$$[\Omega_0^- g](x,v) = \exp\left(-\int_0^\infty \nu(x+v\tau+\tfrac{1}{2}a\tau^2, v+a\tau)d\tau\right) g(x,v),$$

where by definition $\Omega_0^- g = \lim_{t\to+\infty} W_0(-t)S_0(t)g$. Ω_0^- is a positive operator, whose inverse is the bounded positive operator

$$[\Omega_0^-]^{-1}\ h(x,v) = \exp\left(\int_0^\infty \nu(x+v\tau+\tfrac{1}{2}a\tau^2, v+a\tau)d\tau\right) h(x,v).$$

Now, we prove the existence of $\Omega_1^- = \underset{t\to+\infty}{\text{s-lim}}\ S_0(-t)S(t)$. The starting point is the Duhamel formula, which multiplied by $S_0(-t)$ gives with g a nonnegative function

$$S_0(-t)S(t)g = g + \int_0^t S_0(-s)KS(s)g\,ds. \qquad (9.4)$$

First, making use of Assumption 2′, we bound the norm of the operator $S_0(-t)$.

$$\|S_0(-t)g\|_1 \leq \int_{-\infty}^{+\infty}\int_{-\infty}^{+\infty} \exp\Big(\int_{-t}^{0} \nu(x-vs+\tfrac{1}{2}as^2, v-as)ds\Big) g(x+vt+\tfrac{1}{2}at^2, v+at)\,dxdv,$$

and again with the change of variables along the characteristics, $x' = x+vt+\frac{1}{2}at^2$, $v' = v+at$, we obtain for $t \geq 0$,

$$\|S_0(-t)g\|_1 \leq \int_{-\infty}^{+\infty}\int_{-\infty}^{+\infty} \exp\Big(\int_0^{+\infty} \nu(x'-v'\tau+\tfrac{1}{2}a\tau^2, v'-a\tau)d\tau\Big)\, g(x',v')\, dx'dv' \leq$$

$$\leq \int_{-\infty}^{+\infty}\int_{-\infty}^{+\infty} \exp\Big(\frac{1}{a}\int_{-\infty}^{+\infty} \bar{\nu}(w)dw\Big)\, g(x',v')\, dxdv \leq \exp\Big(\frac{M}{a}\Big)\, \|g\|_1 .$$

Thus the operator $S_0(-t)$ is bounded above by $\exp(M/a)$ in $L_1(\mathbf{R}^2, dxdv)$. Now, according to the definition of Ω_1^-, we want to prove that the limit of the right hand side of (9.4) for $t \to +\infty$ exists. Using the above estimate gives

$$\Big\|\int_0^t S_0(-s)KS(s)g\,ds\Big\|_1 \leq \int_0^{+\infty} \|S_0(-s)KS(s)g\|_1 ds \leq \exp\Big(\frac{M}{a}\Big)\int_0^t \|S(s)g\|_\nu ds,$$

where the positive linear operator K satisfies $\|Kf\|_1 = \|f\|_\nu$, when $f \geq 0$, is used. Hence, using (9.3), it is found that

$$\int_0^{+\infty} \|S_0(-s)KS(s)g\|_1\, ds \leq \frac{M}{a}\exp\Big(\frac{2M}{a}\Big)\|g\|_1 < +\infty.$$

These estimates prove the existence Ω_1^- in the strong sense by using the Cook's method in L_1, [30], i.e.

$$\Omega_1^- g = \lim_{t\to+\infty} S_0(-t)S(t)g = g + \int_0^\infty S_0(-s)KS(s)g\,ds,$$

which is a bounded positive operator.

Finally, we define Ω^- by the composition $\Omega_0^-\Omega_1^-$, which satisfies (9.2) and is a bounded positive operator. □

THEOREM 11. If the collision frequency $\nu(x,v)$ satisfies the same assumptions as in Theorem 10, then the limit

$$\Omega^+ = \underset{t\to-\infty}{\text{s-lim}}\ S(-t)W_0(t), \tag{9.5}$$

exists in the strong operator topology of $L_1(\mathbf{R}^2, dxdv)$ and is a bounded positive operator.

Proof. Let $t \leq 0$, we first evaluate $[S_0(-t)W_0(t)g](x,v)$

$$[S_0(-t)h](x,v) = \exp\Big(-\int_0^{-t} \nu(x-vs+\tfrac{1}{2}as^2, v-as)ds\Big)\, h(x+vt+\tfrac{1}{2}at^2, v+at).$$

Putting $h = W_0(t)g$ yields

$$[S_0(-t)\ W_0(t)g]\ (x,v) = \exp\left(-\int_0^{-t} \nu(x-vs+\tfrac{1}{2}as^2, v-as)ds\right) g(x,v).$$

Hence, we have in the strong operator topology of $L_1(\mathbf{R}^2, dxdv)$

$$\Omega_0^+ = \underset{t\to-\infty}{\text{s-lim}}\ S_0(-t)W_0(t),$$

$$[\Omega_0^+ g](x,v) = \exp\left(-\int_0^{+\infty} \nu(x-vs+\tfrac{1}{2}as^2, v-as)ds\right) g(x,v) \geq$$

$$\geq \exp\left(-\frac{1}{a}\int_{-\infty}^{+\infty} \bar{\nu}(\hat{v})\,d\hat{v}\right)g(x,v) = \exp\left(-\frac{M}{a}\right) g(x,v)$$

yielding a positive operator with bounded, positive inverse. Now we consider $S(-t)S_0(t)$. The starting point is

$$S(-t)S_0(t)g = g + \int_t^0 S(-s)KS_0(s)g\,ds, \quad \text{for } t \leq 0$$

obtained by applying $S(-t)$ to both sides of the Duhamel formula. Using the property $\|S(t)g\|_1 = \|g\|_1$, for $g \geq 0$ in $L_1(\mathbf{R}^2, dxdv)$, we obtain

$$\int_{-\infty}^0 \|S(-s)KS_0(s)g\|_1 ds = \int_{-\infty}^0 \|KS_0(s)g\|_1 ds = \int_{-\infty}^0 \|S_0(s)g\|_\nu ds.$$

Moreover, for nonnegative $g \in L_1(\mathbf{R}^2, dxdv)$

$$\int_{-\infty}^0 \|S_0(t)g\|_\nu dt \leq \exp\left(\frac{M}{a}\right)\int_{-\infty}^0\int_{-\infty}^{+\infty}\int_{-\infty}^{+\infty} \nu(x,v)\,g(x-vt+\tfrac{1}{2}at^2, v-at)\,dxdvdt \leq$$

$$\leq \frac{M}{a}\exp\left(\frac{2M}{a}\right)\|g\|_1.$$

Hence in the strong operator topology of $L_1(\mathbf{R}^2, dxdv)$

$$\Omega_1^+ g = \lim_{t\to-\infty} S(-t)S_0(t)g = g + \int_{-\infty}^0 S(-s)KS_0(s)g\,ds,$$

which is a bounded positive operator. Thus $\Omega^+ = \Omega_1^+\Omega_0^+$ is a well-defined bounded positive operator, satisfying (9.5). □

10. <u>TRAVELLING WAVES</u>

In this section, we want to derive some physical conclusions from the existence of the wave operators. Recall that the runaway phenomenon was characterized in Definition 1 by means of the electron velocity distribution. In the following theorem, as a direct consequence of the existence of Ω^-, we prove that the integrabilty of $\bar{\nu}(v)$ is sufficient for the onset of the runaway process.

THEOREM 12. If all the assumptions of Theorem 10 are satisfied, then the physical process modelled by equation (6.1) gives rise to runaway, according to Definition 1.

Proof. We show that there exists a function $\mathcal{F}_\infty=\mathcal{F}_\infty(v)$ such that

$$\lim_{t\to+\infty}\int_{-\infty}^{+\infty}|\mathcal{F}(v+at,t)\ -\ \mathcal{F}_\infty(v)|\ dv=\ 0.$$

First we observe that result of Theorem 10 has to be read as follows. There exists an operator Ω^- such that in the strong operator topology of $L_1(\mathbf{R}^2,dxdv)$

$$\lim_{t\to+\infty}\ \|W_0(-t)S(t)g\ -\ \Omega^- g\|_1\ =\ 0,\ \text{for any}\ g\in L_1(\mathbf{R}^2,dxdv).\ (10.1)$$

More precisely, we can rewrite, for any $g\in L_1(\mathbf{R}^2,dxdv)$

$$\lim_{t\to+\infty}\int_{-\infty}^{+\infty}dv\int_{-\infty}^{+\infty}dx\ |\ [W_0(-t)S(t)g](x,v)\ -\ [\Omega^- g](x,v)\ |\ =\ 0.$$

Let f_0 be the initial condition and let us define $\mathcal{F}_\infty(v)$ = $\int_{-\infty}^{+\infty}[\Omega^- f_0](x,v)\ dx$. Let us denote by $f(x,v,t)$ the solution $S(t)f_0$ and by $\mathcal{F}(v,t)$ the corresponding velocity electron distribution $\int_{-\infty}^{+\infty}f(x,v,t)dx$.

We find that

$$\int_{-\infty}^{+\infty}|\mathcal{F}(v+at,t)\ -\ \mathcal{F}_\infty(v)|\ dv=$$

$$\le\int_{-\infty}^{+\infty}dv\int_{-\infty}^{+\infty}\left|f(x+vt+\tfrac{1}{2}at^2,v+at,t)-[\Omega^- f_0](x,v)\right|dx\ =$$

$$=\int_{-\infty}^{+\infty}dv\int_{-\infty}^{+\infty}dx\ \left|[W_0(-t)S(t)f_0]\ (x,v)\ -\ [\Omega^- f_0](x,v)\right|.$$

Due to the property (10.1), we have the existence of the required limit. When $f_0=f_0(x,v)$ is the initial condition, the limit function $\mathcal{F}_\infty(v)$ is given by $\int_{-\infty}^{+\infty}[\Omega^- f_0](x,v)\ dx$. □

Another physical interpretation of this result can be given, in terms of travelling waves. In velocity space, by a travelling wave with "velocity" a we mean a function of the form $g(v-at)$. Then the previous results can be read in the sense that $\mathcal{F}(v,t)$ behaves like a travelling wave, i.e.

$$\mathcal{F}(v,t)=\int_{-\infty}^{+\infty}[S(t)f_0](x,v)\ dx\ \simeq\ \int_{-\infty}^{+\infty}[\Omega^- f_0](x,v-at)\ dx\ =\ \mathcal{F}_\infty(v-at),$$

as $t\to+\infty$.

PART III

11. THE THREE-DIMENSIONAL VELOCITY MODEL

In this third part, we are concerned with a three-dimensional velocity model for a spatially uniform population of electrons moving within a weakly ionized host medium. In this section and in the following one we will report some results recently obtained by Arlotti, [13], on the time-dependent problem and on the asymptotic behaviour of the solution, in the case of the existence of stationary solutions. In the last section, we will present briefly some work in progress on the long time behaviour in cases where runaway occurs.

Let us consider the linear equation which describes the evolution of the electron distribution $f(\vec{v},t)$, as a function of the velocity $\vec{v}\in\mathbf{R}^3$ and time $t\geq 0$, in a weakly ionized host medium

$$\frac{\partial f}{\partial t}(\vec{v},t)+\vec{a}\cdot\frac{\partial f}{\partial \vec{v}}(\vec{v},t)+\nu(|\vec{v}|)\,f(\vec{v},t) = \int_{\mathbf{R}^3} k(\vec{v},\vec{v}')\,\nu(|\vec{v}'|)\,f(\vec{v}',t)\,d\vec{v}', \quad (11.1)$$

with the initial condition

$$f(\vec{v},0) = f_0(\vec{v}). \quad (11.2)$$

The acceleration field is given by $\vec{a}=-\left|\frac{e}{m}\right|\vec{E}$, where $\vec{E}$ is the applied electrostatic field, time and position independent, and e and m are the charge and the mass of the electron, respectively. Apart from the three-dimensionality, the physical situation is similar to that of previous parts and the meaning of the symbols is quite clear.

Here we do not dwell too long upon the assumptions and the functional formulations, which can be given as in the previous sections.

We choose a reference frame $(O,\vec{i},\vec{j},\vec{k})$ and, with no loss of generality, we assume $\vec{a}=a\vec{k}$, with a constant and positive. The collision frequency $\nu(v)$ is a function of the speed $v=|\vec{v}|$, which in [13] satisfies the following more general assumption, allowing non integrable singularities at the finite:

Assumption 2″. The collision frequency $\nu(v)$ is a Lebesgue measurable nonnegative function of v on $(0,+\infty)$, which is almost everywhere nonzero and

$$J= \left\{v\geq 0:\ \forall h>0\ \text{ess sup}\ \{\nu(v'): v'\in \mathbf{R}^+\cap[v-h,v+h]\}=\infty\right\}$$

is a finite subset of $\mathbf{R}^+$.

Moreover, $k(\vec{v},\vec{v}')d\vec{v}$ is the collision probability, which satisfies the following assumption:

Assumption 3″. The collision kernel $k(\vec{v},\vec{v}')\geq 0$, appearing in the integral operator, is a measurable function from $\mathbf{R}^6$ into $\mathbf{R}^+$, with the property

$$\int_{\mathbf{R}^3} k(\vec{v},\vec{v}')d\vec{v} \equiv 1, \qquad \forall \vec{v}'\in \mathbf{R}^3,$$

and satisfies the reciprocity symmetry $k(-\vec{v},-\vec{v}') = k(\vec{v},\vec{v}')$.

Let us introduce the Banach spaces $L_1(\mathbf{R}^3,d\vec{v})$ and $L_1(\mathbf{R}^3,\nu d\vec{v})$ with the norms

$$\|f\|_1 = \int_{\mathbf{R}^3} |f(\vec{v})|d\vec{v}, \qquad \|f\|_\nu = \int_{\mathbf{R}^3} \nu(|\vec{v}|)\,|f(\vec{v})|d\vec{v}.$$

We denote by T_0, A and K the following operators:

$$\begin{cases} T_0: D(T_0)\subset L_1(\mathbf{R}^3,d\vec{v}) \to L_1(\mathbf{R}^3,d\vec{v}) \\ \qquad T_0 f = -\vec{a}\cdot\dfrac{\partial f}{\partial \vec{v}}, \end{cases}$$

$$\begin{cases} A: L_1(\mathbf{R}^3,\nu d\vec{v})\cap L_1(\mathbf{R}^3,d\vec{v}) \to L_1(\mathbf{R}^3,d\vec{v}) \\ \qquad Af = -\nu(v)f, \end{cases}$$

$$\begin{cases} K: L_1(\mathbf{R}^3,\nu d\vec{v}) \to L_1(\mathbf{R}^3,d\vec{v}) \\ (Kf)(\vec{v}) = \displaystyle\int_{\mathbf{R}^3} k(\vec{v},\vec{v}')\nu(v')f(\vec{v}')\,d\vec{v}', \end{cases}$$

where T_0f is defined in the distributional sense. Then K is a positive linear operator, still satisfying $\|Kf\|_1=\|f\|_\nu$, $f\in L_1(\mathbf{R}^3,\nu d\vec{v})$ and $f\geq 0$.

Now we are able to put problem (11.1)-(11.2) into the following abstract form

$$\frac{df}{dt}(t) = T_0 f(t) + Af(t) + Kf(t), \quad t>0, \tag{11.6}$$

$$\lim_{t\to 0} f(t) = f_0, \tag{11.7}$$

with the usual meaning of the symbols.

The abstract evolution problem (11.6)-(11.7) is investigated in [13], in a more general context; here we summarize the principal results used in the sequel.

LEMMA 13. If Assumption $2''$ is satisfied, then the operator T_0+A is the infinitesimal generator of the strongly continuous semigroup $\{S_0(t);\ t\geq 0\}$, given by

$$[S_0(t)\,g](\vec{v}) = \exp\Big(-\int_0^t \nu(|\vec{v}-\vec{a}s|)\,ds\Big)g(\vec{v}-\vec{a}t), \qquad \text{for any}$$

$g\in L_1(\mathbf{R}^3,d\vec{v})$.

If Assumption $3''$ is also satisfied, then (a closed extension of) T_0+A+K is the infinitesimal generator of the strongly continuous semigroup $\{S(t);\ t\geq 0\}$, on $L_1(\mathbf{R}^3,d\vec{v})$, satisfying

$$S(t)g = S_0(t)g + \int_0^t S(t-s)KS_0(s)g\,ds, \quad \forall g\in D(T_0+A),$$

where $D(T_0+A)=\{f\in L_1(\mathbf{R}^3,d\vec{v})\cap L_1(\mathbf{R}^3,\nu d\vec{v}):\ T_0f\in L_1(\mathbf{R}^3,d\vec{v})\}$.
Moreover, $S(t)$ is a positive semigroup, with $\|S(t)f\|_1\leq\|f\|_1$, $f\in L_1(\mathbf{R}^3,d\vec{v})$ and $f\geq 0$, $t\geq 0$.

12. THE STATIONARY PROBLEM AND THE ASYMPTOTIC BEHAVIOUR

First of all, we examine what type of condition must be satisfied by the collision frequency, when searching for solutions with full physical meaning. As stated above, the problem of relaxation to the steady-state solution is very sensitive to the behaviour of the collision frequency $\nu=\nu(v)$. This time, the mathematical analysis of a necessary condition for existence of stationary solution is slightly different, because in the one-dimensional case, such a necessary condition

$$\int_u^{+\infty} \nu(v')\,dv' = +\infty, \text{ for all positive } u$$

is equivalent to

$$\int_0^{+\infty} \nu(|\vec{v}+\vec{a}s|)ds = +\infty, \text{ for all } \vec{v}\in\mathbf{R}^3. \tag{12.1}$$

as is proved in [18]. Whereas in the three-dimensional case, this equivalence ceases to exist. We can give a simple lemma about the integrability properties of $\nu=\nu(v)$ to clarify the difference with the one-dimensional case. We prove that

$$\int_{-\infty}^{+\infty} \nu\Big(\sqrt{v_x^2+v_y^2+v_z^2}\Big)\,dv_z \tag{12.2}$$

can be infinity for $\forall(v_x,v_y)\in\mathbf{R}^2$ or bounded for a.e. $(v_x,v_y)\in\mathbf{R}^2$, in such a way as to exclude the possibility of having (12.2) bounded for (v_x,v_y) in a set of positive measure different from

$\mathbf{R}^2$.

Let $\rho = \sqrt{v_x^2 + v_y^2}$ and let us put $N(\rho) = \int_{-\infty}^{+\infty} \nu\left(\sqrt{\rho^2 + v_z^2}\right) dv_z$.

PROPOSITION 14. If the function $\nu=\nu(v)$ satisfies Assumption 2 (see Section 2), then only two cases are possible:

i) $N(\rho) = +\infty$ for all $\rho \in [0,+\infty)$;

ii) $N(\rho) < +\infty$ for a.e. $\rho \in [0,+\infty)$.

Proof. When ν, as a function of the speed v, satisfies Assumption 2, we have only two kinds of behaviour: ν is non integrable on $\mathbf{R}$, or alternatively ν is integrable. We show that both these cases correspond to those stated above.

By a simple change of the integration variable, we can write, for $v > \rho$

$$N(\rho) = 2 \int_{\rho}^{v} \nu(v') \frac{v'}{\sqrt{v'^2 - \rho^2}} dv' + 2 \int_{v}^{+\infty} \nu(v') \frac{v'}{\sqrt{v'^2 - \rho^2}} dv'. \tag{12.3}$$

i) Let us suppose $\nu(v)$ non integrable on $\mathbf{R}$; then for all $v \geq 0$

$$\int_{v}^{+\infty} \nu(v')\, dv' = +\infty.$$

For any fixed $\rho \in [0,+\infty)$, we choose $v=\rho$ and case i) is certainly satisfied, because from (12.3)

$$N(\rho) \geq 2 \int_{\rho}^{+\infty} \nu(v')\, dv' = +\infty.$$

ii) Let us suppose $\nu(v)$ integrable on $\mathbf{R}$. Let $n \in \mathbf{N}$ and $\epsilon>0$ be fixed. There exist $M>0$ and a measurable set $A \subseteq [0,n]$, such that $m(A)<\epsilon/3$ and $\nu(v) \leq M$, for all $v \in [0,n] \setminus A$. Because A is measurable, there exists a countable family of disjoint open intervals I_i such that

$$\mathcal{A} = \bigcup_{i=1}^{\infty} I_i \supset A \qquad \text{and} \qquad m(\mathcal{A})<\epsilon.$$

Moreover, according to the fact that the boundary of A has measure zero, $m(\overline{\mathcal{A}})<\epsilon$. We now define B as the set $(0,n) \setminus \overline{\mathcal{A}} \subset [0,n] \setminus A$. Because B is open, for any point ρ fixed in B, there exists $v>\rho$ such that $[\rho,v] \subset B$. Then, using again (12.3),

$$N(\rho) \leq 2M \int_{\rho}^{v} \frac{v'}{\sqrt{v'^2 - \rho^2}} dv' + 2 \frac{v}{\sqrt{v^2 - \rho^2}} \int_{v}^{+\infty} \nu(v')\, dv' < +\infty.$$

Hence, for every $n \in \mathbf{N}$ and $\epsilon>0$, we have $m(\{\rho \in [0,n]: N(\rho)<+\infty\}) > n-\epsilon$. The proposition is completely proved. □

In general $N=N(\rho)$ does not belong to L_∞ with respect to ρ;

$\nu(v)$ could have some singularities thus making $N(\rho)$ infinity on a set of measure zero.

From now on we consider collision frequencies satisfying Assumption $2'$, which allows having a finite number of non integrable singularities. First we can establish the following proposition on the properties of integral (12.1).

PROPOSITION 15. Let $\nu=\nu(v)$ be a function satisfying Assumption $2''$.

If $\int_{\epsilon}^{+\infty}\nu(v)\,dv <+\infty$, for any $\epsilon>0$, then for any $\vec{v}\in\mathbf{R}^3$ such that $|\vec{v}\times\vec{k}|\notin J$ we have $\int_0^{+\infty}\nu(|\vec{v}-\vec{a}t|)dt< +\infty$;

if $\int_{v_1}^{v_2}\nu(v)\,dv <+\infty$, for all $0<v_1<v_2<+\infty$, then for any $\vec{v}\in\mathbf{R}^3$ such that $|\vec{v}\times\vec{k}|\notin J$ and $0<t<+\infty$, we have $\int_0^{t}\nu(|\vec{v}-\vec{a}s|)ds < +\infty$;

finally, if $\int_{u}^{+\infty}\nu(v)\,dv=+\infty$, for any $u\geq 0$, then for any $\vec{v}\in\mathbf{R}^3$ we have $\int_0^{+\infty}\nu(|\vec{v}-\vec{a}t|)dt = +\infty$.

<u>Proof.</u> See Proposition 2 in [13]. □

In this part, we give a definition of stationary solution, relying on the notion of evolution semigroup, in order to give a more direct integral formulation.

Let us define a stationary solution for problem (11.6)-(11.7) a function $f\in L_1(\mathbf{R}^3,d\vec{v})$, such that

$$S(t)f=f, \quad \forall t\geq 0,$$

where $S(t)$ is the strongly continuous semigroup on $L_1(\mathbf{R}^3,d\vec{v})$ generated by T_0+A+K.

We will assume that

$$\int_{u}^{+\infty}\nu(v)\,dv = +\infty, \text{ for all } u\geq 0, \tag{12.4}$$

which is necessary for the existence of stationary solutions, at least when the set J is either empty or equal to {0}. For a proof of this necessity, see Theorem 3 in [13]. Now we are in position to give the following theorem on the integral formulation of the stationary problem, (see Proposition in [13]).

THEOREM 16. Let us suppose that Assumptions $2''$ and $3''$ and the additional property (12.4) are satisfied. A function f belongs to $D(T_0+A)$ and it is a stationary solution to problem

(11.6)-(11.7) if and only if f belongs to $L_1(\mathbf{R}^3,d\vec{v})\cap L_1(\mathbf{R}^3,\nu d\vec{v})$ and is a solution to the integral equation

$$f = \int_0^{+\infty}[S_0(t)Kf]dt.$$

In [13] the stationary problem is completely investigated when the integral kernel of operator K takes the simplified form

$$k(\vec{v},\vec{v}') = h(\vec{v}), \qquad \forall\vec{v}'\in\mathbf{R}^3, \qquad \text{with } h\in L_1(\mathbf{R}^3,d\vec{v}) \text{ and } \|h\|_1=1.$$

In the literature, for a suitable choice of the function h, this is commonly known as the BGK model for the collision operator. Here we limit ourselves to report without proof the theorem on the asymptotic behaviour of the solution.

THEOREM 17. Let us suppose, together with Assumptions $2''$ and $3''$ and the additional property (12.4), that the function h satisfies the properties

i) $h(\vec{v})>0$, for a.e. $\vec{v}\in\mathbf{R}^3$,

ii) $\displaystyle\int_{\mathbf{R}^3\times\mathbf{R}^+} h(\vec{v}) \exp\Big(-\int_0^t \nu(|-\vec{v}-\vec{a}s|)\,ds\Big)d\vec{v}\,dt< +\infty.$

For any f_0 belonging to $L_1(\mathbf{R}^3,d\vec{v})$, the solution $S(t)f_0$ admits limit as $t\to+\infty$, given by $\displaystyle\lim_{t\to+\infty} S(t)f_0 = c_0\chi$, where

$$\chi=\chi(\vec{v})=\int_0^{+\infty}[S_0(t)h](\vec{v})\,dt, \quad \text{and} \quad c_0=\|\chi\|_1^{-1}\int_{\mathbf{R}^3} f_0(\vec{v})d\vec{v}.$$

Requirement (i) can be dropped, thereby obtaining a weaker result, because f_0 must belong to $Y_N=\{f\in L_1(\mathbf{R}^3,d\vec{v}): f=0 \text{ a.e. on } N\}$, with $N=\{\vec{v}\in\mathbf{R}^3: \int_0^{+\infty}[S_0(t)h](\vec{v})\,dt=0\}$, [13]. Moreover we observe that requirement (ii) could be replaced by

$$\int_{v_1}^{v_2}\nu(v)\,dv <+\infty, \text{ for all } 0<v_1<v_2<+\infty.$$

13. <u>LONG TIME BEHAVIOUR IN CASE OF RUNAWAY</u>

In this final section, we are interested in the other main physical situation, in which the collision process is not effective in removing kinetic energy from the electrons so that the distribution function does not relax towards a nonzero steady state distribution. In particular, we will examine the existence of a travelling wave in velocity space in the runaway regime. We have to contradict the necessary condition (12.4), to exclude the existence of stationary solutions. As remarked above, the opposite of (12.4) is not strictly equivalent to the converse of (12.1). In this section, we consider the further hypothesis on $\nu(v)$

$$\int_{-\infty}^{+\infty} \nu(|\vec{v}-\vec{a}s|)ds \le \frac{M}{a} < +\infty, \quad \text{for all } \vec{v}\in\mathbf{R}^3, \tag{13.1}$$

which is sufficient to ensure the existence of the wave operators.

Let $W_0(t)$ be the free streaming group generated by T_0, i.e. $[W_0(t)g](\vec{v}) = g(\vec{v}-at)$, $t\in\mathbf{R}$, and $S(t)$ the full semigroup defined above.

THEOREM 18. If Lemma 13 is true, and in addition (13.1) is satisfied, then the limits

$$\Omega^- = \underset{t\to+\infty}{\text{s-lim}}\; W_0(-t)S(t)$$

$$\Omega^+ = \underset{t\to-\infty}{\text{s-lim}}\; S(-t)W_0(t)$$

exist in the strong operator topology of $L_1(\mathbf{R}^3, d\vec{v})$ and are bounded positive operators.

<u>Proof</u>. Hypothesis (13.1) allows us to compute estimates sufficient to prove the theorem, for which we refer to [31].

More precisely, in the strong operator topology of $L_1(\mathbf{R}^3, d\vec{v})$ we have

$$\Omega_0^- g = \lim_{t\to+\infty} W_0(-t)S_0(t)g, \text{ with } [\Omega_0^- g](\vec{v}) = \exp\left(-\int_0^\infty \nu(|\vec{v}+\vec{a}\tau|)\,d\tau\right)g(\vec{v}),$$

$$\Omega_1^- g = \lim_{t\to+\infty} S_0(-t)S(t)g = g + \int_0^\infty S_0(-s)KS(s)g\,ds,$$

which are bounded positive operators. Ω^- can be defined by $\Omega^- = \Omega_0^-\Omega_1^-$.

Finally, we give a theorem which proves that the solution of the time dependent problem will increasingly resemble a travelling wave in velocity space with "velocity" a as $t\to+\infty$. By a travelling wave (in velocity space) of "velocity" a we mean a function of the form $W_0(t)g$ where g does not depend on t, i.e. a function of the form

$$[W_0(t)g](\vec{v}) = g(\vec{v}-\vec{a}t).$$

In the following theorem, we see how the integrability of the collision frequency affects the asymptotic behaviour of the streaming motion.

THEOREM 19. Following the hypothesis of Theorem 18, we have, for every $g \in L_1(\mathbf{R}^3, d\vec{v})$

$$\lim_{t \to +\infty} \| W_0(-t) S(t) g - \Omega^- g \|_1 = 0.$$

where Ω^- is the wave operator defined above.

Proof. This theorem is a direct consequence of Theorem 18. □

Under the conditions of this theorem we have

$$[S(t)g](\vec{v}) \simeq [\Omega^- g](\vec{v} - \vec{a}t),$$

as $t \to +\infty$, i.e. in the remote future $S(t)g$ looks like a travelling wave in velocity space.

ACKNOWLEDGEMENT

Work performed under the auspices of C.N.R.–G.N.F.M. and of the M.P.I. project "Equations of Evolution." This work began while G.F. was visiting the Department of Mathematical Sciences at University of Delaware, Newark DE 19716 (USA).

This work was first presented as a lecture at the 3rd Workshop on Mathematical Aspects of Fluid and Plasma Dynamics, held in Salice Terme, Italy, September 1988. The author thanks the scientific and organizing commitees for their support.

The author is greatly indebted to S.L. Paveri-Fontana and C.V.M. van der Mee for helpful comments and for having obtained together the results presented in the first part of this lecture. Finally the author thanks L. Arlotti, for her permission to report some results on three-dimensional systems.

REFERENCES

[1] K. Kumar, H.R. Skullerud and R.E. Robson, Kinetic theory of charged particle swarms in neutral gases, Aust. J. Phys. 33, 343-448 (1980)

[2] K. Kumar, The physics of swarms and some basic questions of kinetic theory, Phys. Rep. 112, 319-375 (1984).

[3] H. Dreicer, Electrons and ion runaway in a fully ionized

gas. I, Phys. Rev. 115, 238-249 (1959)

[4] H. Dreicer, Electrons and ion runaway in a fully ionized gas. II, Phys. Rev. 117, 329-342 (1960).

[5] V.V. Parail and O.P. Pogutse, Runaway electrons in a plasma. In: M.A. Leontovich (Ed.), "Reviews of plasma physics," Vol. 11, Consultants Bureau, New York, 1986, pp. 1–63.

[6] M.C. Mackey, Kinetic theory model for ion movement through biological membranes, Biophys. J. 11, 75-95 (1971).

[7] W. Fawcett, A.D. Boardman and S. Swain, Monte Carlo determination of electron transport properties in gallium arsenide, J. Phys. Chem. Solids. 31, 1963-1990 (1970).

[8] M.O. Vassell, Calculation of high-field distribution functions in semiconductors, J. Math. Phys. 11, 408-412 (1970).

[9] N. Corngold and D. Rollins, Diffusion with varying drag. The runaway problem. I, Phys. Fluids 29, 1042-1048 (1986).

[10] N. Corngold and D. Rollins, Diffusion with varying drag. The runaway problem. II, Phys. Fluids 30, 393-398 (1987).

[11] G. Frosali, C.V.M. van der Mee, and S.L. Paveri-Fontana, Conditions for runaway phenomena in the kinetic theory of particle swarms, J. Math. Phys. 30, 1177–1186 (1989).

[12] G. Frosali, and C.V.M. van der Mee, Scattering theory relevant to the linear transport of particle swarms, J. Stat. Phys. 56, 139-148 (1989).

[13] L. Arlotti, On the asymptotic behaviour of electrons in an ionized gas. In: P. Nelson et al. (Eds.), "Transport Theory, Invariant Imbedding, and Integral Equations," Lecture Notes in Pure and Appl. Math., vol. 115, M. Dekker, New York, 1989, pp.81–96.

[14] R. Beals and V. Protopopescu, Abstract time dependent transport equations, J. Math. Anal. Appl. 121, 370-405 (1987).

[15] W. Greenberg, C.V.M. van der Mee and V. Protopopescu, "Boundary value problems in abstract kinetic theory," Basel, Birkhäuser OT 23, 1986.

[16] B. Sherman, The difference-differential equation of electron energy dfistribution in a gas, J. Math. Anal. Appl. 1, 342-354 (1960).

[17] J.L. Delcroix, "Physique des plasmas," Tome 2, Paris, 1966.

[18] G. Cavalleri and S.L. Paveri-Fontana, Drift velocity and

runaway phenomena for electrons in neutral gases, Phys. Rev. A 6, 327-333 (1972).

[19] G.L. Braglia, Motion of electrons and ions in a weakly ionized gas in a field 1. Foundations of the integral theory, Plasmaphysik 3, 147-194 (1980)

[20] M.A. Krasnoselskii, "Positive solutions of operator equations," Groningen, Noordhoff, 1964 = Moscow, Fizmatgiz, 1962 [Russian].

[21] V.C. Boffi and V.G. Molinari, Nonlinear transport problems by factorization of the scattering probability, Nuovo Cimento B 65, 29-44 (1981)

[22] V.C. Boffi, V.G. Molinari, and R. Scardovelli, Kinetic approach to the propagation of electromagnetic waves in a weakly ionized plasma, Nuovo Cimento D, 1, 673-687 (1982).

[23] R. Nagel (Ed.), "One-parameter semigroups of positive operators," Lecture Notes in Mathematics 1184, Berlin, Springer, 1986.

[24] J. Hejtmanek, Scattering theory of the linear Boltzmann operator, Commun. Math. Phys. 43, 109-120 (1975).

[25] B. Simon, Existence of the scattering matrix for the linearized Boltzmann equation, Commun. Math. Phys. 41, 99-108 (1975).

[26] W. Schappacher, Scattering theory for the linear Boltzmann equation, Ber. Math.-Statist. Sekt. Forschungszentrum Graz n. 69, 14 pp. (1976).

[27] J. Voigt, On the existence of the scattering operator for the linear Boltzmann equation, J. Math. Anal. Appl. 58, 541-558 (1977).

[28] C.V.M. van der Mee, Trace theorems and kinetic equations for non divergence free external forces, Applicable Anal., to appear.

[29] F.A. Molinet, Existence, uniqueness and properties of the solutions of the Boltzmann kinetic equation for a weakly ionized gas. I, J. Math. Phys. 18, 984-996 (1977).

[30] T. Kato, "Perturbation theory for linear operators," Springer Verlag, Berlin, 1966.

[31] L. Arlotti, and G. Frosali, Long time behaviour of particle swarms in runaway regime, preprint

FURTHER RESULTS IN THE NONLINEAR STABILITY OF THE MAGNETIC BÉNARD PROBLEM

G.P. Galdi and M. Padula
Dipartimento di Matematica
Via Machiavelli 35, 44100 Ferrara

Introduction

As is known, the Magnetic Bénard Problem (MBP) regards the stability of a motionless horizontal layer of an electrically and thermally conducting fluid heated from below and on which a vertical uniform magnetic field is impressed. Here, one of the most interesting effect is constitued by the inhibiting influence exerted by the magnetic field on the onset of convection (stabilizing effect of the magnetic field). Because of the large interest of this subject, from both experimental and theoretical point of views, such a problem has been widely studied and it results a rapidly advancing area.

Since this related literature is very broad, we shall limit ourselves to recall some of the main theoretical contributions. Precisely, we quote the works of Thomson (1951) and of Chandrasekhar (1981) regarding the linear theory and those of Busse (1975), Proctor & Galloway (1979), Rudraiah (1981), Weiss (1981) concerning the nonlinear approach, based on a formal expansion procedure. We also quote the comprehensive review article of Proctor & Weiss (1982). Further progress has recently been made by Galdi (1985) who considered the full nonlinear stability problem for perturbations whose amplitude is *initially* below a computable constant (conditional stability); this was achieved introducing a new "generalized energy" method, successively reconsidered by Rionero (1988) and Rionero & Mulone (1988). Another step in the field has been recently accomplished by Galdi & Padula (1989), henceforth denoted by GP. Precisely, the authors propose a novel stability approach based upon the concept of operator symmetry. Such a theory works for linear and nonlinear systems as well and provides an algorithm for the construction of the appropriate generalized energy (Liapounov) functional to study the evolution of perturbations.

Regarding the MBP, GP show that the energy functional must contain two "coupling terms" G_1, G_2 which take into accomet the interactions between temperature and magnetic fields (G_1) and kinetic and magnetic field (G_2), see formula (2.1). Moreover, these two terms are just those responsible for the stabilizing influence of the magnetic field. We notice, however, that while G_1 was already considered by Galdi (1985), Rionero (1988), Rionero & Mulone (1988), the term G_2 appears to be new and its magnitude depends on the magnetic Prandtl number P_m. We can, therefore, conclude that P_m plays an important role in the magnetic Bénard problem as a new independent stabilizing parameter.

Such an influence is only qualitatively revealed by Rudraiah (1981). Furthermore, solving a certain variational problem in a suitable class of perturbations, S_1 say, GP show for liquid metals (i.e., $P_m << 1, P_m/P_r << 1, P_r$ = Prandtl number) a strong influence of P_m on the critical Rayleigh number R_e as a function of the Chandrasekhar number Q. In the present paper we solve the same variational problem, again for liquid metals, but in the appropriate class of perturbations $S_2 \subset S_1$. Thus, we find a critical curve $R_E = R_E(Q)$ which improves on that of GP and which essentially coincides with that of linear theory, see Chandrasekhar (1981). Therefore, roughly speaking, our criterion is *necessary as well as sufficient* for nonlinear stability whenever the amplitude of pertubations is *initially* below a suitable bound which can be *quantitatively* determined, see GP. It should also be remarked that, as expected, we again find an influence of P_m on stability in that the critical wave number, unlike linear theory, heavily depends on P_m, see Figure 1.

1. The Magnetic Bénard Problem as a weakly coupled system

Consider an infinite horizontal layer of an electrically and thermally conducting fluid upon which different temperatures T_0, T_1 at the boundary planes and a uniform vertical magnetic field $\mathbf{H} = H\mathbf{k}$ are imposed, where $\mathbf{k}$ is the upword vertical direction, As is well known, denoting by $\{0; x, y, z\}$ a coordinate frame with $\mathbf{k}$ = grad z, the static state $S_0 = \{\mathbf{v} = 0 \;\; P = const \;\; T = (T_1 - T_0)z \;\; \mathbf{H} = H\mathbf{k}\}$ satisfies the basic equations in the Bousinnesq approximation, cfr. e.g. Chandrasekhar (1981) for every values of the physical parameters entering the equations.

The generic perturbation $\{\mathbf{u}, p, \theta, \mathbf{h}\}$ to S_0 is thus solution of the following non dimensional system.

$$\frac{\partial \mathbf{u}}{\partial t} = -\nabla x\, \nabla x \mathbf{u} + R\,\theta \mathbf{k} + Q\mathbf{h}_{,z} - \nabla p - \mathbf{u}\cdot\nabla\mathbf{u} - P_m\, \mathbf{h}\cdot\nabla\mathbf{h}$$

$$\nabla\cdot\mathbf{u} = 0 \qquad \nabla\cdot\mathbf{h} = 0 \tag{1.1}$$

$$P_r \frac{\partial\theta}{\partial t} = \Delta\theta + Rw - P_r\,\mathbf{u}\cdot\nabla\theta$$

$$P_m \frac{\partial \mathbf{h}}{\partial t} = -\nabla x \nabla x \mathbf{h} + Q\,\mathbf{u}_{,z} - P_m\,(\mathbf{u}\cdot\nabla\mathbf{h} - \mathbf{h}\cdot\nabla\mathbf{u})$$

where $R = \dfrac{\alpha g \beta d^4}{\chi \nu_c}$, $Q = \dfrac{\mu_m H^2 d^2}{4\pi\rho\nu_c\eta_m}$ are the Rayleigh and Chandrasekhar numbers, ν_c the kinematical viscosity, ρ the material density, η_m the resistivity, α the coefficient of volume expansion, μ_m the magnetic permeability g the gravity, $\beta = T_1 - T_0 > 0$ and T_1 the temperature at the bottom, χ the thermal diffusivity and d the depth of the layer. Moreover, $P_r = \nu_c/\chi$, $P_m = \nu_c/\eta_m$ the thermal and magnetic Prandtl numbers. Finally, we set, z $\equiv \partial/\partial z$. To (1.1) we append the following conditions at the planes z = 0,1

$$(\mathbf{u} x \mathbf{k})_{,z} = 0 \quad \mathbf{u} \cdot \mathbf{k} = 0 \quad \theta = 0 \quad \mathbf{h} x \mathbf{k} = 0 \qquad \text{at } z = 0, 1 \tag{1.2}$$

These conditions for free bounding surfaces are appropriate in modelling stellar convection, see Chandrasekhar (1981), Roberts (1967), Peckover & Weiss (1972). Rigid motions and uniform magnetic fields are excluded by the global conditions

$$\int_C (\mathbf{u} x \mathbf{k})\, dx = \int_C \mathbf{h} \cdot \mathbf{k}\, dx = 0 \tag{1.3}$$

where C denotes the "periodicity cell" $C = \{(x, y, z) \in (0, 2\pi/a_1] x \ (0, 2\pi/a_2] x (0, 1)\}$ $a_1, a_2 > 0$.

In order to study the evolution in time for the perturbation (u, p), with $u \equiv (\mathbf{u}, \theta, \mathbf{h})$, it turns useful to introduce some functional spaces. Let $X = \left[C^8(\bar{C})\right]^7$ where $C^8(\bar{C})$ is the usual space of functions which are continuous up to their 8-th derivatives in $\bar{C}$, and $L^2(C)$ the usual Lebesgue space. Moreover, $J^*(C)$ is the subspace of solenoidal vector functions of $[L^2(C)]^3$ and $J(C)$ is the subspace of $J^*(C)$ with $\mathbf{u} \cdot \mathbf{k} = 0$ at $z = 0, 1$. Furthermore, we set $H = J(C) x L(C) x J^*(C)$, and introduce the following operators on H.

$$Bu = \begin{pmatrix} \mathbf{u} \\ P_r \theta \\ P_m \mathbf{h} \end{pmatrix} \qquad Au = \begin{pmatrix} -\nabla x \nabla x \mathbf{u} \\ \Delta \theta \\ -\nabla x \nabla x \mathbf{h} \end{pmatrix}$$

$$Su = \begin{pmatrix} \Pi(\theta \mathbf{k}) \\ w \\ 0 \end{pmatrix} \qquad Mu = \begin{pmatrix} \mathbf{h}_{,z} \\ 0 \\ \mathbf{u}_{,z} \end{pmatrix} \tag{1.4}$$

$$Nu = \begin{pmatrix} -\Pi(\mathbf{u} \cdot \nabla \mathbf{u} - P_m \mathbf{h} \cdot \nabla \mathbf{h} \\ -P_r \mathbf{u} \cdot \nabla \theta \\ -P_m(\mathbf{u} \cdot \nabla \mathbf{h} - \mathbf{h} \cdot \nabla \mathbf{u}) \end{pmatrix}$$

here B, A, S, M are linear operators on H while Nu is non linear, moreover Π is the projection operator of $L^2(C)$ onto $J(C)$. In view of (1.4) system (1.1) can be put in the abstract form in $X \cap H$

$$u_{,t} = Au + RSu + QMa + Nu \tag{1.5}$$

In order to incorporate the boundary conditions (1.2), (1.3) into the problem we denote by

$Y(A) = \left\{u \in H : \mathbf{u}, \theta, \mathbf{h} \textit{ verify } (1.2) \text{ and } (1.3) ; \int_C \left\{|\nabla x \nabla x\, \mathbf{u}|^2 + |\Delta\theta|^2 + |\nabla x \nabla x \mathbf{h}|^2\right\} dx < +\infty\right\}$

$Y(M) = \left\{u \in H : \mathbf{h} x \mathbf{k} \textit{ verifies } (1.2)_4 ; \int_C \left\{|\mathbf{u}_{,z}|^2 + |\mathbf{h}_{,z}|^2\right\} dx < +\infty\right\}$ the domains of definition of A and M, respectively.

Finally, we let R_E be the energy parameter:

$$R_E^{-1} = \max \frac{2\int_C \varphi\psi \cdot \mathbf{k}\, dx}{\int_C (|\nabla\psi|^2 + |\nabla\phi|^2 + |\nabla\chi|^2)\, dx} \tag{1.6}$$

where the maximum is taken for $(\psi, \phi, \chi) \in Y(A)$. We recall that R_E^{-1} coincides with the maximum taken for $\chi = 0$ (Rionero (1968)) and therefore $R_E = 25.641$, Joseph (1966).

As shown in GP problem (1.5) constitutes a *weakly conpled system*. For reader's convenience we recall the basic assumption of a weakly coupled system, see GP.

A system of the form (1.5) is said to define a weakly conpled system if S and M are symmetric [1] and skew symmetric, respectively and $R_E < +\infty$. Moreover, the following "interaction properties" between B, A, S, M are supposed to hold:

$$\begin{aligned} AB &= BA \\ B^{-1/2} S B^{-1/2} &= K_1 S \\ B^{-1/2} M B^{-1/2} &= K_2 M \end{aligned} \tag{H.1}$$

with K_1, K_2 constants. Moreover, $Y(M) \supseteq Y(A)$ and

$$\begin{aligned} Mu \in Y(A) \quad & \text{for } any \quad u \in Y(M) \cap X \\ Au \in Y(M) \quad & \text{for } any \quad u \in Y(A) \cap X \\ MAu = AMu \quad & \text{for } any \quad u \in Y(A) \cap X \end{aligned} \tag{H.2}$$

Finally, denoting by (,) the scalar product in H and setting $D(u) = -(Au, u)$ we require for some $\gamma_1 > 0$

$$\begin{aligned} (Su, u) &\le \gamma_1 [D(u) D(M_{u)}]^{1/2)} \\ (SMu, Mu) &= 0 \end{aligned} \tag{H.3}$$

Systems of type (1.5) verifyng the set of hypotheses (H.1)-(H.3) are called *weakly coupled*. In GP it is proved that (1.4),(1.5) with $Y(A), Y(M)$ previously specified, define a weakly coupled system with $K_1 = P_r^{-1/2}$, $K_2 = P_m^{-1/2}$ and $\gamma_1 = 2/\pi^3$.

Referring the reader to GP for a discussion on the meaning of hypotheses (H.1)-(H.3) in the general case, here we may wish to comment some of the them for the case at hand.First of all (H.1) is identically satisfied if $B = I$ (the identity operator), while, in general, $B \neq I$ may lead to some new physical effect on stability. In the magnetic Bénard problem, the presence of B is intimately related to the dissipation mechanisms (i.e., to the viscosity and to the resistivity), and its importance will be widely clarified in section 3 below. Assumption (H.3)$_1$ requires that the "destabilizing" part S must be dominated by the new dissipation $D(Mu)$ due just to the presence of the magnetic field.

Finally (H.3)$_3$ is related to the coupling between S and M and constitues a basic assumption on the mechanism of the interaction. In the case at hand, it requires that there is no direct interaction between temperature and magnetic fields. In this sense the system is weakly coupled; actually there is a destabilizing influence due to the direct action of the temperature on the thermally conducting fluid (simple Bénard problem) and a stabilizing effect due to the action of the magnetic field on the electrically conducting fluid. Thus, the magnetic field acts on the temperature only *indirectly* (weakly) through the kinetic field.

---

(1) The hypothesis of symmetry for S by no means needed and it can be dropped, see GP, Galdi (1989), Coscia & Padula (1989).

2. Choice of the generalized energy functional and variational formulation of stability.

Once it has been recognized that the MBP defines a weakly coupled system, we may perform as in GP the energy analysis in order to show the stabilizing effect of the magnetic field. To this end, to select the energy functional we first compute the coupling functionals G_1 and G_2 which are responsible for the stabilizing effect. As shown in GP they are given by

$$G_1(u) \equiv (S\,u\,, M\,u)$$
$$G_2(u) \equiv (M\,u\,, B\,u)$$

so that we have (with $h = \mathbf{h} \cdot \mathbf{k}$)

$$G_1 = \int_C \Pi\,(\theta\,\mathbf{k} \cdot \mathbf{h}_{,z}\,dx = -\int_C h\,\theta_{,z}\,dx$$
$$G_2 = (1 - P_m)\int_C \mathbf{u} \cdot \mathbf{h}_{,z}\,dx \qquad (2.1)$$

Unlike the heuristic generalized energy functional introduced by Galdi (1985) and reconsidered by Rionero & Mulone (1988) where only $G_1(u)$ was included into the energy functional, in GP for the first time also $G_2(u)$ was considered as a completely new coupling term, whose contribution is particulary linked to P_m. Also, in GP was performed a first nonlinear stability analysis including the coupling term $G_2(u)$. Precisely it was shown that in the case $P_m \to 0$ (liquid metals) the stability region $R = R(Q)$ was markedly enlarged compared to the earlier results of Galdi (1985), Rionero & Mulone (1988). In view of (2.1) and the work of GP, the energy functional to be choosen here is given by

$$E(t) = \frac{1}{2}\int_C \left\{ \mathbf{u}^2 + P_r\,\theta^2 + P_m\,h^2 + \lambda_2\,(|\mathbf{u}_{,z}|^2 + p_m\,|\mathbf{h}_{,z}|^2) + 2\lambda(P_m\,P_r)^{1/2}\,\theta\,h_{,z} + 2\lambda_3\,\mathbf{u} \cdot \mathbf{h}_{,z} \right\} dx \qquad (2.2)$$

$$\lambda_3 = \lambda_1\sqrt{P_m}/(1-P_m)\,,\ \lambda_2 = \lambda Q/(R\sqrt{p}) \qquad \lambda = \xi Q/(R\sqrt{p}) \qquad \xi \in (0\,,1)$$

where $p = P_m/P_r$. In order that E be positive definite we now select $\lambda_3 = \xi\,\nu\,\pi\,\sqrt{P_m}\,Q^2/pR^2$, $|\nu| \in (0\,,\sqrt{1-\xi})$. We remark that for liquid metals the energy functional E is the best one for the study of the *linear* stability of S_0, namely, when we set $N\,u \equiv 0$. Actually we shall prove that in the case $P_m << 1$, the stability curve $R_E = R_E(Q)$ we shall compute essentially coincides with that $R_L = R_L(Q)$ of linear stability of Chandrasekhar (1981). However, what is more important the condition $R < R_E = R_E(Q)$ *ensures also rigorous nonlinear conditional asymptotic stability.* As shown in GP, this latter objective can be reached by using the more "regular" energy

$$\mathrm{E} = E(t) + \eta\,E_1(t) \qquad (2.3)$$

with η suitable positive parameter and

$$E_1(t) = \frac{1}{2}\int_C \left\{ \mathbf{u}^2 + P_r\,\theta^2 + P_m\,\mathbf{h}^2 + |\nabla \mathbf{u}|^2 + P_r\,|\nabla\theta|^2 + P_m\,|\nabla \mathbf{h}|^2 \right\} dx$$

fictitious energy first introduced by Galdi (1985). Actually in GP it is proved (for an arbitrary weakly coupled system) that conditions ensuring $dE/dt \le 0$ along the solutions to the *linearized* equation (obtained by formally setting $N \equiv 0$ in (1.5)) imply the *nonlinear* exponential asymptotic conditional stability of the basic state S_0 with respect to the energy functional E. We emphasize one more time that at $t=0$ is supposed to be less than a suitable *computable* constant, see GP. Thus, evaluating dE/dt along (1.4), (1.5) with $N \equiv 0$ we deduce

$$\frac{d}{dt}E = \Im(u) - \wp(u) \tag{2.4}$$

where

$$\Im(u) \equiv \int_C \{ 2R\,\theta\, w + \lambda R\sqrt{p}\,w h_{,z} + \lambda_3 R\,\theta h_{,z} - \lambda_3(1+P_m^{-1})\nabla \mathbf{u} : \nabla \mathbf{h}_{,z} - \lambda\left(\frac{1+p}{\sqrt{p}}\right)\nabla h_{,z}\cdot\nabla\theta$$

$$- \lambda_3 Q\left(\frac{|\mathbf{u}_{,z}|^2}{P_m} - |\mathbf{h}_{,z}|^2\right)\}\,dx$$

$$\wp(u) \equiv \int_C \left\{ \frac{\lambda Q}{R\sqrt{p}}\left(|\nabla \mathbf{u}_{,z}|^2 + |\nabla \mathbf{h}_{,z}|^2\right) + |\nabla \mathbf{u}|^2 + |\nabla\theta|^2 + |\nabla \mathbf{h}|^2 \right\} dx \tag{2.5}$$

We now set

$$m \equiv \max_S \frac{\Im(u)}{\wp(u)} \tag{2.6}$$

where $S \subseteq Y(A) \cap X$ is the class of (regular) solutions to (1.1)-(1.3). Condition $dE/dt \le 0$ for all $t \ge 0$ will be verified if and only if $m < 1$. In order to obtain a quantitative stability criterion, in the next section we shall solve the variational problem (2.6).

3. On the resolution of the characteristic problem and numerical results in the case of liquid metals.

As previously remarked, in order to derive sufficient conditions for $m < 1$ we are led to solve the E Euler-Lagrange equations associated to (2.6). The characteristic eigenvalue problem connected with (2.6) is

$$m\,\Delta \mathbf{u} - m\frac{\lambda Q}{R\sqrt{p}}\Delta \mathbf{u}_{,zz} + R\left(\theta + \frac{\sqrt{p}}{2}h_{,z}\right)\mathbf{k} + \frac{\lambda_3}{2}\left(1+\frac{1}{P_m}\right)\Delta \mathbf{h}_{,z} + \lambda_3\frac{Q}{P_m}\mathbf{u}_{,zz} - \nabla X = 0$$

$$\nabla\cdot\mathbf{u} = 0$$

$$m\,\Delta\theta + R\,w + \frac{\lambda(1+p)}{2\sqrt{p}}\,\Delta\,\mathbf{h}_{,z} + \frac{\lambda_3 R}{2}\,\mathbf{h}_{,z} = 0$$

(3.1)

$$m\,\Delta\,\mathbf{h} - m\,\frac{\lambda Q}{R\sqrt{p}}\,\Delta\,\mathbf{h}_{,zz} - \frac{\lambda(1+p)}{2\sqrt{p}}\,\Delta\theta_{,z}\,\mathbf{k} - \frac{\lambda R\sqrt{p}}{2}\,w_{,z}\,\mathbf{k} - \frac{\lambda_3 R}{2}\,\theta_{,z}\,\mathbf{k} - \frac{\lambda_3}{2}\,(1+\frac{1}{P_m})\,\Delta\mathbf{u}_{,z}$$

$$-\lambda_3\,Q\,\mathbf{h}_{,zz} - \nabla\psi = 0$$

$$\nabla\cdot\mathbf{h} = 0$$

From the assumed periodicity of perturbations, we may follow the standard normal modes analysis and write $w = W(z)\,e^{2i\,(a_1 x + a_2 y)}$, ... , etc. Operating with $\nabla x\,\nabla x$ into $(3.1)_{1,4}$ and multiplying by $\mathbf{k}$ we get rid of the Lagrange parameters X and ψ. Moreover, we solve the remaining three scalar differential equations in w, θ, h.

After a somewhat tedious but straightforward manipulation, we obtain the following equation in w (analogous equations in θ, h can be deduced)

$$L_1 W = -\,L_2 W \tag{3.2}$$

with $(a^2 = a_1^2 + a_2^2)$

$$L_1 \equiv \Big\{m(D^2-a^2)^2\,[\,(D^2-a^2) - \frac{\lambda Q}{R\sqrt{p}}\,D^2(D^2-a^2) - \lambda_3 Q D^2] - \frac{a^2D^2}{m}\Big[\frac{\lambda(1+p)}{2\sqrt{p}}\,(D^2-a^2) + \frac{\lambda_3 R}{2}\Big]^2\Big\}x$$
$$x\left\{\frac{R^2}{m}\,a^2 + (D^2-a^2)^2\,D^2\,\frac{\lambda_3 Q}{P_m} - m(D^2-a^2)^2\,x\,[\frac{\lambda Q}{R\sqrt{p}}\,(D^2-a^2)D^2 - (D^2-a^2)]\right\}$$

(3.2)

$$L_2 \equiv D^2\,\Big\{(D^2-a^2)\,[-\frac{\lambda R\sqrt{p}}{2}\,a^2 + \frac{\lambda_3}{2}(1+\frac{1}{P_m})(D^2-a^2)^2] + \frac{R}{m}a^2[\frac{\lambda(1+p)}{2\sqrt{p}}(D^2-a^2) + \frac{\lambda_3 R}{2}]\Big\}\,x$$
$$x\,\Big\{(D^2-a^2)[-\frac{\lambda R\sqrt{p}}{2}\,a^2 + \frac{\lambda_3}{2}(1+\frac{1}{P_m})(D^2-a^2)^2] + \frac{Ra^2}{m}\,[\frac{\lambda(1+p)}{2\sqrt{p}}(D^2-a^2) + \frac{\lambda_3 R}{2}]\Big\}.$$

System (3.2) is an ordinary differential equation of 14-th order in $W = W(z)$ the appropriate boundary conditions to (3.2) are deduced from the system (1.1)-(1.3) and from (3.1). Reasoning as in GP we can show

$$D^{2l}W = 0 \qquad \text{at } z = 0,1 \tag{3.3}$$

for all $l = 0,1, \ldots$ This allows us to conclude that the appropriate solutions to (3.2) (3.3) are of the form $W(z) = H\ \sin\,r\pi z\,,\ r = 1,2,\ldots$

Substituting these functions in (3.2) we obtain a fouth order algebraic equation in m .

$$g\,(m\,,R) \equiv am^4 + bm^3 + cm^2 + dm + e = 0.$$

In order to provide sufficient conditions on R for m to be less than one we proceed as follows.

I. We solve the equation in R^*

$$g(1, R^*) = 0 \tag{3.4}$$

As we shall see below (3.4) represents a 3-rd order equation in $(R^*)^2$. Under suitable assumption on ξ, ν we can prove that there exists only one positive real solution

$$R^* = R^*(r, a, \xi, \nu, P_m, Q, p)$$

We set

$$R_E \equiv \max_{\xi,\nu} \min_{r,a} R^* \equiv R_E(P_m, Q, p)$$

II. We prove for all $R<R_E$

$$g(1, R) > 0$$

$$\frac{\partial g}{\partial m}(m, R) > 0 \qquad m>1$$

$$\frac{\partial^2 g}{\partial m^2}(m, R) > 0 \qquad m>1$$

III. $R<R_E \Rightarrow m<1$

The computation of R_E will be given below in the case $p<<1$ and $P_m<<1$ (liquid metals). The proof of II is given in GP when the variational problem (2.6) is solved in the wider class S_1 where the solenoidality of **h** is not required. Such a proof works equally well in this case and here is not reproduced.

Point III is obvious because from II it follows $g(m,R)>0$ $m>1$, $R<R_E$ and therefore for $R<R_E$ the only points m such that $g(m,R)=0$ must satisfy the condition $m<1$.

The remaining part of this section is devoted to solve the equation $g(1,R^*)=0$, i.e.,

$$z^3 + Az^2 + Bz + C = 0 \tag{3.5}$$

where $z=(R^*)^2$.

Setting $\xi = 4p\delta \dfrac{(1+x)(1-\sigma)}{x(1+p-\sigma)^2}$, $x = \dfrac{a^2}{r^2\pi^2}$, $\sigma = \dfrac{Q\sqrt{P_m}\,\nu}{\pi^2(1+x)}$, $\omega = \dfrac{1}{P_m}$, $L = \dfrac{r^2\pi^2(1+x)^2\omega^2}{4Q^2(1-\delta)}$

we have

$$A = -r^4\pi^4\frac{(1+x)^3}{x} + \frac{\xi}{p}x\,(1-\sigma)\,Q^2\,[\,(1+x)\,(1-\delta) + \frac{\xi}{4p}\eta\,(1-\sigma)]$$

$$B = -r^6\pi^6\frac{(1+x)^3}{x}Q^2\frac{\xi}{p}\left[1+\sigma\omega + (1-\sigma)\,(1-\delta) + \frac{2x}{1+x}\sigma\omega\,(1-\sigma)\frac{\xi}{4p}\right] \tag{3.6}$$

$$C = -r^8\pi^8\,Q^4\,\frac{\xi^2}{p^2}\frac{(1+x)^3}{x}(1-\delta)\,[\,(1+\sigma\omega)\,(1-\sigma) - \sigma^2 M\,]$$

In order for (3.5) to have only one positive real solution z we require $\delta<½$. It can be proved, see GP, that the minimum of z in r is taken for $r=1$. In the limits

$$P_m << 1,\ p << 1$$

the coefficients A, B, C in (3.6) of the cubic equation (3.5) become

$$A(t,\delta,x) = -\pi^4\frac{(1+x)^3}{x} + 4\pi^2 Q^2\delta\frac{(1+x)}{x}$$

$$B\,(t,\delta,x) = -\pi^6\frac{(1+x)^3}{x}Q^2\frac{\xi}{p}\left[2 + t\sigma_+\omega - \delta\,(1+2t\sigma_+\omega)\right] \tag{3.7}$$

$$C\,(t,\delta,x) = -\pi^8\frac{(1+x)^3}{x}Q^4\,(\frac{\xi}{p})^2\,(1-\delta)\,(t-1)\left[\sigma_+^2\ (L+\omega)t+1\right]$$

and $\sigma_+ = \left\{\omega + \sqrt{\omega^2+4(L+\omega)}\right\}/\left[2(L+\omega)\right]$ solution of $C(\sigma_+) = 0$, $\sigma = t\sigma_+$, $t \in (0,1)$..

Obviously, the sign of the coefficients A,B,C and their magnitude will govern the magnitude of $z^+ = z^+\,(x,\delta,t)$ where z^+ denotes the only positive root of (3.5). The maximization of z^+ in is a difficult points. In order to maximize z^+ in terms of δ, t we notice that $\sigma_+\ 1 \Leftrightarrow L>1$. Then, one possible choice for σ can be $\sigma=\sigma_+$, i.e. $t=1$ if $L>1$.

The requirement $L>1$ implies the existence of a value Q_c for Q, with $Q_c = \dfrac{\pi^2\omega^2}{4\,(1-\delta)}$ below which, i.e. $Q<Q_c$, the solution $y^+ = z^+\,(1,\delta,x)$ solves also the quadratic equation

$$y^2 + Ay + B_0 \equiv q\,(y) = 0 \tag{3.8}$$

and B_0 is the value of B evaluated at $t=1$. The solution y^+ is still a function of δ and x and it should be maximized in δ and minimized in x. Let us denote by $y_0 = y^+\,(\bar{\delta},x)$ the generic positive solution of (3.8) where $\delta = .15$. It is easy to convince oneself that setting

$$C(z,\delta,t) = z^3 + A(\delta)\,z^2 + B\,(\delta,t)z + C(\delta,t)\,,$$

$$q(y,\eth) = y^2 + A(\eth)y + B(\eth,1)$$

it holds

$$C(y^+,\eth,t) > 0, \qquad \text{for all } t \in (0,1) \tag{3.10}$$

Actually we have

$$C(y^+,\eth,t) = \left[B(\eth,t) - B(\eth,1)\right]y^+ + C(\eth,t)$$

and for $\omega >> 1$ the condition (3.10) is equivalent to

$$Q < \frac{y^+(1-2\eth)}{4\pi^2\eth(1-\eth)} \tag{3.11}$$

Therefore, whenever (3.11) is satisfied relation (3.10) is true, this fact implies, in particular, that the positive zero y^+ of $q(y,\eth)$ is greater than the positive zero z^+ of $C(z,\eth,t)$. As a consequence, we must choose the value y^+ as a critical Rayleigh number whenever (3.11) holds . Numerical evidence reveals that (3.11) is always satisfied when $Q<Q_c$. In this respect, we observe that for a typical liquid metal like mercury where $\omega = 7.10^6$ the critical Chandrasekhar number is $Q_c^2 \cong 1.5 \cdot 10^{14}$, so that up to these values of Q^2 the critical Rayleigh number R_E is simply obtained by solving the quadratic equation (3.8) after minimization over $x>0$. On the other hand, in general, for $Q>Q_c$, R_E is likewise obtained by solving the cubic equation (3.5). Values of R_E and of the corresponding critical value a_c of x (i.e. that which, at fixed Q, minimizes R_E) as function of Q for $P_m = 10^{-2}$ and $P_m = 6 \cdot 10^{-8}$ are given in Table 1. A comparison is also given with the critical Rayleigh number R_L computed by Chandrasekhar (1981) using linear theory. Notice that linear theory gives the same value of R_L and a_c for any $P_m<1$. On the other hand, from our generalized non-linear energy analysis it comes out that a_c heavily depends on P_m , see also Fig. 1. Finally, it is apparent that R_E and R_L are very close.

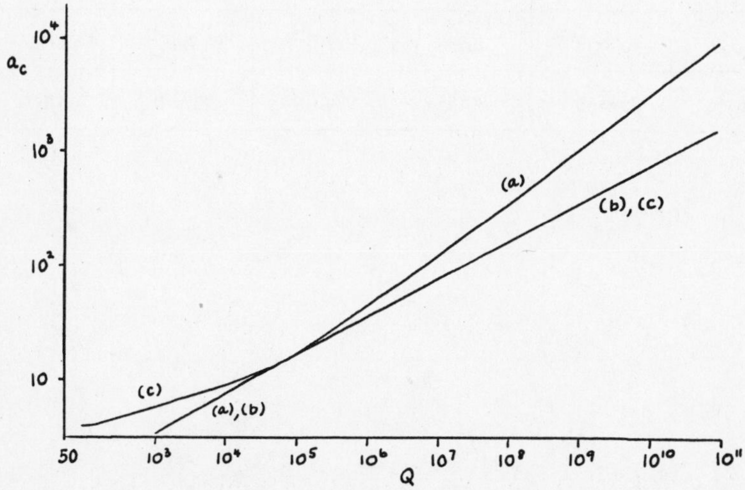

Fig. 1 - Variation of the critical amplitude a_c with Q:

(a) Present theory, Pm = 10^{-2}

(b) Present theory, Pm = 6.10^{-8}

(c) Linear Theory

$\omega = 100$ $\omega = 1.5 \cdot 10^7$

Q	a_c	R_E	a_c	R_E	R_L	a_c (linear)
10	.6	909	.6	896	923	2.5
100	1.3	2559	1.4	25×10^2	2653	3.7
10^3	3.3	14.6×10^3	3	14×10^3	15×10^3	5.7
10^4	7.5	11.5×10^4	7.3	11×10^4	12×10^4	8.6
10^5	17.5	10.4×10^5	16	10×10^5	10.8×10^5	16.7
10^6	43	99.4×10^5	36	95.9×10^5	10×10^6	36.5
10^7	110	97.7×10^6	77	93.8×10^6	10×10^7	79.2
10^8	346	97.1×10^7	168	92.8×10^7	99.5×10^7	171
10^9	1096	96.9×10^8	361	92.4×10^8	99.1×10^8	369
10^{10}	3466	96.8×10^9	779	92.1×10^9	98.8×10^9	796
10^{11}	10961	96.8×10^{10}	1678	92×10^{10}	98.8×10^{10}	1717
10^{12}	34663	96.8×10^{11}	3615	92×10^{11}	98.7×10^{11}	3699
10^{13}	109616	96.8×10^{12}	7788	92×10^{12}	98.7×10^{12}	7971
10^{14}	346639	96.8×10^{13}	16780	92×10^{13}	98.7×10^{13}	17174
10^{15}	1096166	96.8×10^{14}	36155	92×10^{14}	908.7×10^{14}	37001

Table 1

BIBLIOGRAFY

Busse, F.H., 1975, J. Fluid Mech., **71**, 193.

Chandrasekhar, S., 1981, Hydrodynamic and Hydromagnetic Stability, Dover Publ. Inc., N.Y.

Coscia, V. & M. Padula, 1989, Proc. of the 4-th Int. Conf. on Computational Methods and Experimental Measurements, Capri, Italy, Carlomagno, G.M. & C.A. Brebbia Eds., Srpinger Verlag Pubbl.

Galdi, G.P., 1985, Arch. Ratl. Mech. Anal., **87**, 167.

Galdi, G.P., 1989, Proc. of the 4-th Int. Conf. on Computational Methods and Experimental Measurements, Capri, Italy, Carlomagno, G.M. & C.A. Brebbia Eds, Srpinger Verlag Pubbl.

Galdi, G.P. & M. Padula, 1989, Arch. Ratl. Mech. Anal. in the Press.

Joseph, D.D., 1966, Arch. Ratl. Mech. Anal, **22**, 163.

Peckover, R.S. & N.O. Weiss, 1972, Comp. Phys. Comm., **4**, 339.

Proctor, M.R.E. & D.J. Galloway, 1979, J. Fluid Mech., **90**, 273.

Proctor, M.R.E. & N.O. Weiss, 1982, Rep. Progr. Phys., **45**, 1317.

Rionero, S., 1988, Proc. Meeting "Energy Stability and Convection", Galdi, G.P. & B. Straughan Eds, Pitman Research Notes in Mathematics, **168**.

Rionero, S. & G. Mulone, 1988, Arch. Ratl. Mech. Anal., **103**, 347.

Roberts, P.H., 1967, An Introduction to Magnetohydrodynamics, Longmans: London.

Rudraiah, N., 1981, Publ. Astron. Soc. Japan, **33**, 721.

Thomson, W.B., 1951, Phil. Mag., **42**, 1417.

Weiss, N.O., 1981, J. Fluid Mech., **108**, 247.

PARTICLE TRANSPORT IN NONHOMOGENEOUS MEDIA

by François Golse
Centre de Mathématiques Appliquées
Ecole Normale Supérieure; Paris

§1. Introduction.

This article is meant to be an expository account of some current research on the homogenization of kinetic equations. I shall mainly announce and describe some results in this direction. A few results presented here were announced in [G1].

The subject matter of this talk being the kinetic theory of particles interacting with a background composite material, one has first to discuss the various scales involved. We shall mainly distinguish between three different orders of magnitude: L will be the size (diameter e.g.) of the domain in which the interaction particle/matter takes place; ε will be the characteristic length of the inhomogeneities (that is, the average size of the inhomogeneities, or the average distance between inhomogeneities); finally, d will be the mean free path of particles between two interactions. The cases where $\varepsilon << d$ and $d << \varepsilon$ lead to very different homogenization procedures and results. In the second case ($d << \varepsilon$), the kinetic equation can first be approximated by a diffusion equation describing the evolution of local equilibria: therefore, on is naturally led to homogenize an elliptic or parabolic equation. This is indeed well-known since the work of Bensoussan-Lions-Papanicolaou (see for example [BLP1]) and the results of Murat-Tartar on H-convergence (see [T1,T2,MT]). Various references are devoted to the specific problem described here (kinetic theory with $d << \varepsilon$): [Se1,BLP2] (this last reference considers the crossover case $d \simeq \varepsilon << L$) and [Ar] for Radiative Transfer problems.

Up to our lnowledge, the first case ($\varepsilon << d$) has received little attention; indeed, one is led, in this case, to homogenize a Transport equation which is first order hyperbolic. Part of the mathematical interest in this kind of problems (besides its applications to kinetic theory) comes from a) its connection with areas of mathematics different from PDEs (see [G2] for an application to dynamical systems) b) the fact that it is very different from the homogenization of second order elliptic equations. The only results that I know for this last case ($\varepsilon << d$) were communicated by Dave Levermore: they are based on a master equation approach and concern special probabilistic laws of distribution of the inhomogeneities: see [LPSW,L,LW]. Our approach is different, and based on deterministic techniques. Before coming to the description of our results, let us comment on the very basic facts of homogenization. The physical properties of the composite material are represented in the governig PDE by coefficients that are strongly oscillating with respect to the space variable. There are essentially three ways to represent these oscillations: i) the coefficients are stochastic processes along each particle path, with prescribed distribution; ii) the coefficients are of the form $A(x,x/\varepsilon)$, where $x/\varepsilon=y$ lies in some torus; iii) the coefficients are of the form $A_\varepsilon(x)$, where A_ε converges in some weak L^p topology. Case i) is referred to as stochastic homogenization (see the work of Keller and Papanicolaou, for example); case ii) is referred to as the periodic case (see [BLP1] for an application to second order, mainly elliptic equations). Case iii) contains case ii) and case i) when the correlation lengths go to zero; this was the approach chosen by Murat and Tartar. I shall mainly follow approach iii) in this talk, and explain why (and how) approach ii) can be deceiptive for kinetic equations.

Outline of the article

§2. Velocity Averaging and Product Lemma.

At the basis of each homogenization result lies the fact that certain quantities, say X and Y, are independent, so that E(XY)=E(X)E(Y) (where E represents the average). In our deterministic approach, this means the continuity of certain products with respect to the weak topology involved: $X_\varepsilon \to X$, $Y_\varepsilon \to Y$ and $X_\varepsilon Y_\varepsilon \to XY$ weakly.

Example: For 2nd order elliptic equations of the form $-\mathrm{div}(A_\varepsilon(x)\nabla u_\varepsilon)=f$, $u_\varepsilon \in H^1{}_0(\Omega)$, $f \in H^{-1}(\Omega)$, A_ε in a bounded set of $L^\infty(\Omega)$ and uniformly elliptic, one may take $X_\varepsilon = A_\varepsilon \nabla u_\varepsilon$ and $Y_\varepsilon = \nabla u_\varepsilon$ (special case of the compensated compactness of Murat and Tartar: see [MT]).

We are going to exhibit the same kind of property for kinetic equations; it will be based on the velocity averaging method. The velocity averaging method is a series of results, the prototype of which being the compactness result of [GPS], extended in the form of various regularity reults in [GLPS]. Here is, essentially, the main result of [GPS,GLPS].

Velocity Averaging Theorem. Let μ be a positive bounded measure on $\mathbb{R}^N$ such that

(H) $\mu(\{v \in \mathbb{R}^N \text{ s.t. } |v.\xi| < \alpha|\xi|\}) \to 0$ (resp. $\leq C\alpha^\gamma$)

when α goes to zero, uniformly in ξ. Let $u_\varepsilon = u_\varepsilon(x,v)$ be a bounded family of

$L^2(dx\mu(dv))$ such that the family $v.\partial_x u_\varepsilon$ is bounded in $L^2(dx\mu(dv))$. Then the family of velocity averages $\tilde{u}_\varepsilon=\int u_\varepsilon(x,v)\mu(dv)$ lies in a compact subset of $L^2(dx)$ (resp. in a bounded subset of $H^s{}_{loc}(x)$ with $s=\inf(1,\gamma/2)$).

NB. We shall denote the velocity averages indifferently by $\tilde{u}$, or $<u>$ in the case of more cumbersome expressions.

Various extensions of this result have been given: see [dPL1,GPo] for kinetic equations with force fields (electric or electromagnetic or of more general nature), [G3,GéGo,Gé1] for applications to the theory of pseudo differential operators, and [Gé2] for its connections with compensated compactness. It was also used as one of the main ingredients in the proof of global existence of renormalized solutions of the Boltzmann equation (see [dPL2]).

In order to tackle homogenization problems in kinetic theory, we shall moreover have to use the following lemma as a supplement to the velocity averaging theorem.

Product Lemma. Let $f_\varepsilon=f_\varepsilon(x,v)\to f$ in $L^2{}_{loc}$ and $g_\varepsilon=g_\varepsilon(x,v)\to g$ in $L^2{}_{loc}$ (with $(x,v)\in\mathbb{R}^N\times\mathbb{R}^N$) Assume that

a) for all $\chi=\chi(x,v)\in C_0^\infty$, the velocity averages $<\chi f_\varepsilon>$ lie in some compact subset of L^2;

b) for all $\theta=\theta(x,v)\in C_0^\infty$, $\iint|\theta g_\varepsilon(x,v+h)-\theta g_\varepsilon(x,v)|^2dxdv\to 0$ when $|h|\to 0$, uniformly in ε.

Then $f_\varepsilon g_\varepsilon\to fg$ in the sense of distributions.

The idea for the proof of this product lemma is given in [G1] (in fact, [G1] contains a slightly more general result); various extensions of this are discussed with the help of 2-microlocalization in [Gé2,GéGo].

§3. Homogenization of Neutronics Problems.

We shall be interested in the two following homogenization problems for the Transport equation of Neutronics.

Problem A. How to compute the effective absorption and scattering cross-sections for a composite material irradiated with neutrons?

Problem B. How to test whether a composite material is critical, subcritical or supercritical?

These questions are of obvious interest in Nuclear Engineering (think, for example, of the irradiation of a turbulent mixture of water and vapour in nuclear reactor).

Here (as in all the paper) we limit ourselves to the first scaling evoked in the introduction ($d>>\varepsilon$): thus we set d and L to fixed values and let ε go to zero.

The oscillations of the absorption cross-section $\sigma_\varepsilon(x,v)$ and of scattering cross-section $k_\varepsilon(x,v,v')$ correspond to the transition from one inhomogeneity to the next one: it is therefore a purely spatial phenomenon. Therefore, we can always assume that

$0<C_1<\sigma_\varepsilon(x,v);\ k_\varepsilon(x,v,v')<C_2<\infty$

$$\sigma_\varepsilon \to \sigma \text{ and } k_\varepsilon \to k \text{ in } L^\infty \text{ weak }* \tag{3.1}$$

σ_ε satisfies assumption b) of the Product Lemma in §2; and for all $\theta=\theta(x,v,v')$ in C_0^∞

$$\iiint |k_\varepsilon \theta(x,v,v'+h') - k_\varepsilon \theta(x,v,v')| dx dv dv' \to 0 \text{ when } |h'| \to 0, \text{ uniformly in } \varepsilon. \tag{3.2}$$

Also, the set of admissible positions is X, bounded regular convex open set of $\mathbb{R}^N$, and the set of admissible velocities is some centered closed ball (of finite radius) denoted by V. We denote by Γ_- the set

$\Gamma_-=\{(x,v)\in\partial X\times V \text{ s.t. } v.N_x<0\}$

where n_x is the unit outward normal vector at point $x\in\partial X$. Let us first look at

Problem A: Consider the Cauchy problem for the Transport equation

$$\partial_t u_\varepsilon + v.\nabla_x u_\varepsilon + \sigma_\varepsilon u_\varepsilon - K_\varepsilon u_\varepsilon = 0,\ x\in X,\ v\in V; \tag{3.3}$$

$$u_{\varepsilon|\Gamma_-} = g(x,v)$$

$$u_{\varepsilon|t=0} = f(x,v)$$

(with the notation $(K_\varepsilon\varphi)(x,v)=\int_V k_\varepsilon(x,v,v')\varphi(x,v')dv'$) and assume that $0\leqslant f,g\leqslant C$.

From the maximum principle for the Transport equation, we deduce the estimate

$$0\leqslant u_\varepsilon \leqslant Ce^{(C_2|V|-C_1)t} \tag{3.4}$$

By using the Velocity Averaging theorem and the Product Lemma of §2, we deduce the following statement, which is a first answer to Problem A.

Theorem 3.1. The solution u_ε of (3.3) converges to u for the weak * topology of $L^\infty([0,T]\times X\times V)$ (for all T>0), where u is the solution of

$$\partial_t u + v.\nabla_x u + \sigma u - Ku = 0,\ x\in X,\ v\in V; \tag{3.3}$$

$$u_{|\Gamma_-} = g(x,v)$$

$$u_{|t=0} = f(x,v)$$

with the notation $(K\varphi)(x,v)=\int_V k(x,v,v')\varphi(x,v')dv'$.

This result means that the equivalent absorption cross-section and the equivalent scattering cross-section are determined (at the first order of approximation) by taking the weak-limit of the corresponding microscopic cross-sections. It correspond exactly to what is called "atomic mix" or "ensemble average" in [LPSW] (Theorem 3.1 contains the result of [LPSW]- established for particular distributions of inhomogeneities- in the limit where the correlation length goes to zero). See [G4] for the proof.

Problem B. Define the operators

$$A_\varepsilon = v.\nabla_x + \sigma_\varepsilon - K_\varepsilon,\quad A = v.\nabla_x + \sigma - K$$

with domain

$$D=D(A_\varepsilon)=D(A)=\{\varphi\in L^2(X\times V) \text{ s.t. } v.\nabla_x\varphi\in L^2 \text{ and } \varphi_{|\Gamma_-}=0\}$$

and the corresponding semigroups

$S_\varepsilon(t)=e^{-tA_\varepsilon}$, $S(t)=e^{-tA}$, $t\geq 0$.

The principal eigenvalues of A_ε and A are defined as

$\omega_\varepsilon=\inf sp(A_\varepsilon)=\text{type } S_\varepsilon$, $\omega=\inf sp(A)=\text{type } S$

The corresponding eigenspaces are one dimensional and have positive eigenfunctions as generators. These considerations are crucial for the initial question: X is said critical (resp. subcritical, supercritical) for A_ε if $\omega_\varepsilon=0$ (resp. $\omega_\varepsilon>0$, $\omega_\varepsilon<0$). All this is but classical material and can be found in any basic textbook of Nuclear Engineering (see [GPS, Se2, DL] for a résumé). To fix the notations, we define ψ_ε and ψ by

$A_\varepsilon\psi_\varepsilon=\omega_\varepsilon\psi_\varepsilon$, $\|\psi_\varepsilon\|_{L^2}=1$, $\psi_\varepsilon\geq 0$; $A\psi=\omega\psi$, $\|\psi\|_{L^2}=1$, $\psi\geq 0$;

Theorem 3.2. When ε goes to zero, ω_ε converges to ω; and any subsequence of ψ_ε converges to a function of the form $\lambda\psi$ in L^2 weak *, with $\lambda>0$ (of course, λ depends a priori on the subsequence).

See [G4] for the proof. The meaning of this theorem for practical purposes is that the procedure of "ensemble averaging" also works to answer problem B, that is to test whether a composite fissile material is critical, subcritical or supercritical. One of the critical points here is that the size of the domain should be chosen large enough to make sure that ω_ε stay uniformly above the region $\{\text{Re } z\leq -\inf \sigma_\varepsilon\}$.

§4. Homogenization of Radiative Transfer Problems.

The method described in §2 allows to homogenize various different models arising in Radiative Transfer. We refer the interested reader to [G1,GSe] for more information about these models. Here, we shall concentrate on a model which may appear very crude to describe Radiative Transfer phenomena, but which indicates quite well the mathematical difficulties to be solved. Again, we choose the first scaling of §1: we set d and L to specified values, and let ε go to zero. Our model is concerned with "grey" atmospheres; that is, the opacity is assumed to be independent of the frequency is averaged out of the equations. Here again, the set of admissible positions is X as in §2, and the set of admissible velocities is $V=S^{N-1}$. The radiative intensity (averaged with respect to the frequency) is denoted by $u=u(t,x,v)$, and the fourth power of the temperature (which is the quantity of interest according to the Stefan-Boltzmann law) is denoted by $\varphi=\varphi(t,x)$ -we assume of course local thermodynamical equilibrium. The opacity is denoted by $\sigma_\varepsilon=\sigma_\varepsilon(x)$ and assumed to be independent of φ (that is of the temperature). We keep the notation $\tilde{f}$ for the velocity average $\tilde{f}=\int_{S^{N-1}} f(v)dv/|S^{N-1}|$. Now our models reads:

$$\partial_t u_\varepsilon+v.\nabla_x u_\varepsilon+\sigma_\varepsilon(u_\varepsilon-\varphi_\varepsilon)=0,\ x\in X,\ v\in S^{N-1};$$

$$\partial_t\varphi_\varepsilon+\sigma_\varepsilon(\varphi_\varepsilon-\tilde{u}_\varepsilon)=0,\ x\in X; \tag{4.1}$$

$$u_{\varepsilon|\Gamma\_}=g(x,v)$$

$$u_{\varepsilon|t=0}=f(x,v),\ \varphi_{\varepsilon|t=0}=h(x)$$

with the assumption

$$0\leqslant f,g,h,\sigma_\varepsilon\leqslant C \tag{4.2}$$

Observe the (strange) fact that the internal energy is proportional to φ, with a coefficient independent of x; this fact allows to solve the second equation of (4.1) for φ_ε in terms of $\tilde{u}_\varepsilon$, and is crucial for our purposes, unfortunately (see [GSe] for more realistic assumptions). Strange as it may appear, this simplest model will lead to a quite intricate homogenized problem: in particular, it is impossible to define an equivalent opacity, as we shall see.

Observe that the second equation in (4.1) is an ODE where x is only a parameter, so that there is no hope to know the limit of $\sigma_\varepsilon\varphi|\varepsilon$ by using the techniques of §2. The result is summarized in the following

Lemma 4.1. Assume that $0 \leq a_\varepsilon(x) \leq C$ and that $a_\varepsilon \to \nu$ in the sense of L. C. Young's generalized functions (that is, $\nu=(\nu_x)_{x\in X}$ is a measurable family of probability measures on $\mathbb{R}$ such that $F(a_\varepsilon) \to 1$: $x \to <\nu_x;F> = \int F \nu_x(d\lambda)$ in L^∞ weak *). Consider now the ODE

$$dy_\varepsilon/dt + a_\varepsilon(x) y_\varepsilon = 0,\ y_{\varepsilon|t=0} = f(x) \tag{4.3}$$

Then, for all $T>0$, $y_\varepsilon \to y$ in $L^\infty([0,T]\times X)$ weak *, where y denotes the solution of the following delayed ODE

$$dy/dt + a(x)y = \int_0^t K(t-s,x)y(s,x)ds,\ y_{|t=0} = f(x) \tag{4.4}$$

with

$$a = \text{w-lim } a_\varepsilon = <\nu_x;\lambda>;\ LK(p,x) = <\nu_x;p+\lambda> - <\nu_x;(p+\lambda)^{-1}>^{-1} \tag{4.5}$$

(L denoting the Laplace transform in t, p the Laplace variable corresponding to t).

See [G4] for the proof. Actually, this result is by no means original; similar calculations, not using the formalism of Young generalized functions, can be found in [LPSW]. After the publication of [G1,G3] I became aware of the reference [T2] where a similar lemma is established for second order ODEs. With lemma 4.1, the homogenization of (4.1) follows the lines sketched in §2. We assume that $\sigma_\varepsilon \to \nu$ in the sense of Young, and define

$\bar{\sigma} = <\nu;\lambda> = \text{w-lim } \sigma_\varepsilon$; $K=K(t,x)$ such that

$$LK(p,x) = <\nu;p+\lambda> - <\nu;(p+\lambda)^{-1}>^{-1} \tag{4.6}$$

Our homogenization result is contained in the following

Theorem 4.2. When ε goes to zero, $(u_\varepsilon,\varphi_\varepsilon) \to (u,\varphi)$ in $L^\infty([0,T]\times X\times S^{N-1}) \times L^\infty([0,T]\times X)$ weak *, where (u,φ) is the solution of

$$\partial_t u+v\,\nabla_x u+\sigma(u-\varphi)=-\int_0^t K(t-s,x)(\varphi-\tilde{u})(s,x)ds,\ x\in X,\ v\in S^{N-1};$$

$$\partial_t\varphi+\sigma(\varphi-\tilde{u})=\int_0^t K(t-s,x)(\varphi-\tilde{u})(s,x)ds,\ x\in X; \tag{4.7}$$

$$u_{\varepsilon\Gamma\_}=g(x,v)$$

$$u_{|t=0}=f(x,v),\ \varphi_{|t=0}=h(x)$$

In [G1], the following particular example was treated: σ_ε was an periodically oscillating function taking two constant values on the torus, σ_1 and σ_2, with proportion χ. Therefore, $\nu_x=(1-\chi)\delta_{\sigma_1(x)}+\chi\delta_{\sigma_2(x)}$ with $0\leqslant\chi\leqslant 1$, $\bar{\sigma}=(1-\chi)\sigma_1+\chi\sigma_2$, $<\sigma>^{-1}=(1-\chi)\sigma_1{}^{-1}+\chi\sigma_2{}^{-1}$, Var $\sigma=$Var $\nu=<\nu;\lambda^2>-<\nu;\lambda>^2$, and $\sigma^*=$Var $\sigma/(\bar{\sigma}-<\sigma>)$; it was found that

$$K(t,x)=\text{Var }\sigma.e^{-t\sigma^*} \tag{4.8}$$

Notice that χ could very well depend on x in the example above (in order to simplify the exposition, the result of kG1] was presented with σ_ε uniformly periodic).

The message of the theorem above, if any, is that there is not such a thing as an equivalent opacity, even for the simplest possible Radiative Transfer model containing a coupling between the radiation field and the matter: the form of the homogenized problem is different from that of (4.1) due to the appearance of the delay terms in the rhs. of (4.7). See [G4] for the proof, as well as [GSe] for more information about homogenization and Radiative Transfer.

§5. How Asymptotic Expansions Fail.

Asymptotic expansions in the case of periodic inhomogeneities are useful whenever we are concerned with multiscale phenomena (see next paragraph). It is however an amazing feature of the Transport equation that they fail to give the first order approximation of the equivalent cross-sections. We shall see the reason for it in an instant, and emphasize the difference with similar calculations for second order elliptic equations, as they are treated in [BLP1].

We proceed by applying the techniques of [BLP1] to the following equation (that is, a Transport equation with isotropic scattering):

$$\partial_t u_\varepsilon + v.\nabla_x u_\varepsilon + \Sigma(x,x/\varepsilon)(u_\varepsilon - \tilde{u}_\varepsilon) = 0, \ x\in X, \ v\in V \tag{5.1}$$

with the notation $\tilde{\varphi}=\int_V \varphi(v)dv/|V|$, and the assumption $\sigma_\varepsilon(x)=\Sigma(x,x/\varepsilon)$, where the periodic variable $x/\varepsilon=y$ lies in the N dimensional torus Y. For the moment, we do not bother with initial or boundary conditions: even at a formal level, we shall see that expansions cannot be pushed further than the order zero. As in [BLP1], we seek the solution as an expansion in the form

$$u_\varepsilon(t,x,v)=u_0(t,x,y,v)+\varepsilon u_1(t,x,y,v)+\ldots_{|y=x/\varepsilon} \tag{5.2}$$

that we plug into (6.1). Identification of successive powers of ε leads to

$\varepsilon^{-1}/\ v.\nabla_y u_0=0$

$\varepsilon^{-0}/\ \partial_t u_0 + v.\nabla_x u_0 + \Sigma(x,y)(u_0-\tilde{u}_0) + v.\nabla_y u_1 = 0$

and so on. The first equation is solved by taking $u_0=u_0(t,x,v)$ independent of y. The second equation splits as

$$\partial_t u_0 + v.\nabla_x u_0 + \overline{\Sigma}(x)(u_0-\tilde{u}_0)=0 \tag{5.3}$$

$$v.\nabla_y u_1 = -\{\Sigma(x,y)-\overline{\Sigma}(x)\}(u_0-\tilde{u}_0) \tag{5.4}$$

where the notation $\overline{\varphi}$ stands for $\int_Y \varphi(y)dy/|Y|$. In other words, we have to introduce the corrector term χ defined by

$$v.\nabla_y \chi = -\{\Sigma(x,y)-\overline{\Sigma}(x)\},\ \overline{\chi}=0 \tag{5.5}$$

which is an equation on the torus Y, with (t,x) as parameters, and write u_1 as

$u_1(t,x,y,v)=\overline{u}_1(t,x,v)-\chi(t,x,y)(u_0-\tilde{u}_0)(t,x,v)$

(observe that the y-dependence of u_1 would be due only to the presence of χ). Then $\overline{u}_1$ would be determined by looking at the first order terms

$\varepsilon^{1}/\ \partial_t u_1 + v.\nabla_x u_1 + \Sigma(x,y)(u_1-\tilde{u}_1) + v.\nabla_y u_2 = 0$

which, after averaging over y, would give

$\partial_t\bar{u}_1+v.\nabla_x\bar{u}_1+\Sigma(x,y)(\bar{u}_1-\int_V\bar{u}_1\,dv/|V|)-\int_Y\Sigma\chi dy/|Y|(u_0-\tilde{u}_0)=$

$\partial_t\bar{u}_1+v.\nabla_x\bar{u}_1+\Sigma(x,y)(\bar{u}_1-\int_V\bar{u}_1\,dv/|V|)=0$

(because, formally, $\int_Y\Sigma\chi dy=\int_Y(\Sigma-\Sigma)\chi dy=(1/2)\int_Y v.\nabla_y\chi^2dy=0$)

and the expansion would be constructed recursively in this way, by repeatedly solving equations in the form of (5.5).

There is but one major flaw in this procedure, which is that (5.5) cannot be solved, at least globally in v, in any L^p space. Indeed, trying to represent the solution of (5.5) by a Fourier series leads to

$$\chi=\sum_{k\neq 0}\hat{a}(k,v)e^{ik.v}/ik.v,\text{ with } a=-\{\Sigma-\Sigma\} \qquad (5.6)$$

which obviously does not converge because of the possible apearance of small divisors v.k. There exists an abundant literature about (5.6) (which is referred to in the bibliography of the survey article by J.-B. Bost on KAM theory [Bo]); indeed, to solve repeatedly (5.5), and to estimate the size of χ in terms of the measure of the set of vs for which (5.5) is solved lies at the heart of KAM theory (in its many disguises), ever since the first contribution by Kolmogorov himself in 1954.

Since Siegel, the idea used in these articles starts as follows. Assume that v satisfies a Diophantine condition with parameters (α,τ), as follows:

$$|v.k|\geq\alpha|k|^{-\tau}\text{ for all } k\in\mathbb{Z}^N\backslash 0\ (\alpha>0) \qquad (5.7)$$

then, for such a v, it is possible to solve (5.5), which results in an estimate of the form

$$\|\chi(.,v)\|_{C^m}\leq A_s\|a(.,v)\|_{C^{m+s+N+\tau}} \qquad (5.8)$$

Observe that the smaller τ is, the bigger the set of excluded vs, and the better (interms of loss of derivatives) estimate (5.8) is.

We shall not dwell any longer on these considerations, but propose instead an alternate approach to solve (5.5) <u>once.</u> Instead of solving exactly (5.5) by excluding some undesirable vs, the idea is to solve, for all vs, an equation close to (5.5). To be more precise, we solve

$$\lambda\chi+v.\nabla_y\chi=-\{\Sigma(x,y)-\Sigma(x)\},\ \lambda\neq 0 \qquad (5.9)$$

Assume that

$\mathrm{meas}(\{v \text{ s.t. } |v.k|\leqslant\alpha|k|\})\leqslant C\alpha^{\gamma},\ 0<\gamma<1;$ (5.10)

then, solving (5.9) leads to the estimate

$\|\chi\|_{L^2}\leqslant C\|\hat{a}\|_{L^\infty}\mathrm{Log}(1/\lambda)$

(the proof is very much in the same spirit as the one of the Velocity Averaging theorem; see [G2]). The idea is then to plug the solution fo (5.9) in the expansions above, instead of trying to solve (5.5) for separate vs, and to finally adjust λ interms of ε. The problem is that such a strategy can be used only once. To cut short, the output of all this is as follows. Going back to §3 and to Problem A, we can give a precision to theorem 3.1 in the case of periodic inhomogeneities.

Theorem 6.1. [G4] With the notations above, if Σ is regular enough (say C^{N+1}), then

$\|u_\varepsilon(t,.,.)-u(t,.,.)\|_{L^1(X\times V)}=O(\varepsilon\mathrm{Log}(1/\varepsilon))$ (5.12)

locally uniformly in t.

The striking fact in this result is that this lowest order perturbation estimate for a Transport equation leads to something comparable to Neistadt's result concerning the averaging of perturbations of integrable dynamical systems (quoted in [Arn]) although the Transport equation above is not a Liouville equation for any dynamical system (at least a deterministic one). In fact, equation (5.1) has indeed something to do with a random dynamical system (because of the presence of the scattering term that is closely linked to a Poisson process: see [BLP2] for a precise formulation of all this). A natural question would be to investigate whether Neistadt's result is still true for such random dynamical systems.

§6. Homogenization vs Diffusion Approximation.

In this paragraph, we go back to the problem making the distinction between the two different scalings evoked in §1. Thus we set $d=\varepsilon^{\alpha}$, and we shall have to distinguish in the homogenization process according to whether $\alpha>1$, or $\alpha<1$, or even $\alpha=1$. In order to give a sense to this assumption, we shall only consider periodic inhomogeneities.

We assume the reader well acquainted with the idea of approximating Transport equations by Diffusion equations (see [BSS, DL]). Moreover, for the sake of simplicity, we shall consider here only the simplest model leading to a diffusion approximation, that is, the one of isotropic scattering. We keep the notations of §5. The problem of interest, posed in the whole space to avoid boundary layers, reads as follows

$$\partial_t u_\varepsilon+\varepsilon^{-\alpha} v\, \nabla_x u_\varepsilon+\varepsilon^{-2\alpha}\Sigma(x,x/\varepsilon)(u_\varepsilon-\tilde{u}_\varepsilon)=0,\ x\in X,\ v\in V \tag{6.1}$$

$$u_{\varepsilon|t=0}=f(x)$$

with the assumption

$$0\leqslant f\leqslant C,\ 0<C_1\leqslant\Sigma\leqslant C_2<\infty,\ \int f(x)dx\leqslant C \tag{6.2}$$

With the restriction that $\alpha<1/2$, the result is summarized in the following

Theorem 5.1. [G4] Assume that $\alpha<1/2$ and that $\Sigma\in C^\infty$. When ε goes to zero, $u_\varepsilon\to u=u(t,x)$ in any $L^p([0,T]\times\mathbb{R}^N\times V)$ (for any $T>0$ and $p<\infty$) where u is the solution of

$$\partial_t u-\nabla_x.\{(D/\Sigma(x))\nabla_x u\}=0,\ x\in\mathbb{R}^N, \tag{6.3}$$

where $D=\mathrm{Tr}\{\int_V v\otimes v dv/|V|\}$

The restriction that $\alpha<1/2$ comes from the previous paragraph: the diffusion approximation requires an asymptotic ansatz up to order 2 (being a second order equation); we therefore have to make sure that we can construct simultaneously the expansion corresponding to the diffusion and the one corresponding to the homogenization (which stops at order one, as we saw in §5). We have therefore to assume that the speed of ergodization corresponding to the homogenization procedure is fast enough compared to the one corresponding to the diffusion approximation.

Let us comment on the differences with the cases $\alpha>1$ and $\alpha=1$. The case $\alpha>1$ was treated by Sentis in [Se1]: he found a diffusion equation like (6.3), but with a diffusion coefficient equal to $D.H\text{-}\lim \Sigma(x,x/\varepsilon)^{-1}$. (Usually $\Sigma(x)^{-1} \neq H\text{-}\lim \Sigma(x,x/\varepsilon)^{-1}$ except in the case where x is one dimensional: see [BLP1]). The case $\alpha=1$ was treated in [BLP2] and leads to a diffusion equation again, but with still another diffusion coefficient. Indeed, the diffusion coefficient in the case $\alpha=1$ is given by

$$\iint_{Y\times V} (I-\mathcal{A})^{-1}\mathcal{A}(v/\Sigma(x,y))\otimes v\,dv\,dy/|V||Y|$$

where $\mathcal{A}$ is the operator

$$\varphi \to \mathcal{A}\varphi = (I+\Sigma(x,y)^{-1}v.\nabla_y)^{-1}\tilde{\varphi}$$

(Notice that $\mathcal{A}$ is a compact operator parametrized by x, according to the Velocity Averaging theorem). As a conclusion to this paragraph, it is clear that the three different scalings lead to distinct homogenized diffusion limits; this fact accounts for the distinction as regards the scales that we introduced in §1.

§7. Nonhomogeneous Boundaries.

The effect of rough surfaces on a gas lies at the heart of some modern technologies (see the example of computer hard disks). The only results I can quote in this direction are due to Babovski [Bab] and to Babovski-Bardos-Platkowski [BBP]. I wish to report, in this last paragraph, a result obtained in collaboration with P. Gérard. We do not treat exactly rough surfaces, but composite surfaces with oscillating accomodation coefficients. Such surfaces can be obtained by various processes: molecular beam epitaxy, for example, or weaved materials. At the mathematical level, the technical difficulty comes from the fact that the homogenization bears on traces of the solution, for which more interior regularity is needed. Here is the simplest possible model.

The surface is the boundary of X^c, a bounded regular convex closed set of $\mathbb{R}^N$; the set of positions (for the particles) being X. We look at a stationary transport equation of the form

$$\lambda u+v.\nabla_X u=\gamma\tilde{u}+f \tag{7.1}$$

with $\lambda>\gamma>0$, $f=f(x)\in L^1\cap L^\infty$. The boundary condition is

$$u_\varepsilon(x,v)=(1-a_\varepsilon(x))u_\varepsilon(x,R_Xv)+a_\varepsilon(x)\beta(x)\int_{v'.n_x<0}|v'.n_X|u_\varepsilon(x,v')dv',\ v.n_X>0 \tag{7.2}$$

where $R_Xv=v-2v.n_Xn_X$ and $\beta(x)=(\int_{v'.n_x>0}v'.n_X\, u_\varepsilon(x,v')dv')^{-1}$, so that

$$\int_V v'.n_X u_\varepsilon(x,v')dv'=0 \text{ for all } x\in\partial X \tag{7.3}$$

The accomodation coefficient a_ε is such that

$$a_\varepsilon\to a \text{ in } L^\infty \text{ weak *},\ 0<a_\varepsilon<1 \tag{7.4}$$

Our results are summarized in the

Theorem 7.1. [GéGo] Assume that $f(x)\in L^1\cap L^\infty$. When ε goes to zero, $u_\varepsilon\to u$ in $L^\infty(X\times V)$ weak *, where u is the solution of (7.1) supplemented with (7.2) with a_ε replaced by its weak limit a.

The result for such a result is that products of weakly convergent sequences affect only the density of particles having interacted once with the surface, then being repelled in X and regularized by the effect of the scattering proportional to the velocity average. We then use the velocity averaging theorem in a bootstrap argument, plus the product lemma on the traces.

(*)Acknowledgements. This work has been supported by the C.E.A. (Centre d'Etudes de Limeil-Valenton) under contract n°2312/457.

I am also grateful to Pr. R. Sentis for useful discussions concerning the matter presented in §§3,4,6 of this paper.

References

[Arn]: V. Arnold: Chapitres supplémentaires de la théorie des équations différentielles ordinaires; Mir.

[Ar]: M. Artola: Homogénéisation d'un problème relatif à l'équation de la chaleur non linéaire unidimensionnelle; Note CEA (CELV 15/09/83).

[Bab]: H. Babovski: Math. Methods in the Appl. Sci..

[BBP]: H. Babovski, C. Bardos, T. Platkowski: Diffusion Approximation for a Knudsen Gas in a Thin Domain with Accomodation on the Boundary; preprint.

[BSS]: C. Bardos, R. Santos, R. Sentis: Diffusion Approximation and Computation of the Critical Size of a Transport Operator; Trans; of the AMS, 284, (1984), 617-649.

[BLP1]: A. Bensoussan, J.-L. Lions, G; Papanicolaou: Asymptotic Study of Periodic Structures; North Holland.

[BLP2]: A. Bensoussan, J.-L. Lions, G; Papanicolaou: Boundary Layers and Homogenization of Periodic Structures; J. Publ. RIMS Kyoto Univ., 15, (1979), 53-157.

[Bo]: J.-B. Bost: Tores invariants des systèmes dynamiques hamiltoniens; Sém. Bourbaki 639; (1984-85).

[DL]: R. Dautray, J.-L. Lions: Analyse Mathématique et Calcul Numérique pour les Sciences et les Techniques, Vol. III; Masson.

[dPL1]: R. Di Perna, P. -L. Lions: Global Weak Solutions of the Vlasov-Maxwell System; Comm. on Pure and Appl. Math., to appear.

[dPL2]: R. Di Perna, P. -L. Lions: On the Cauchy Problem for the Boltzmann Equation: Global Existence and Weak Stability Results; Ann. of Math., to appear.

[Gé1]: P. Gérard: Moyennisation et régularité 2-microlocale; preprint ENS LMENS 89-01.

[Gé2]: P. Gérard: Compacité par compensation et régularité 2 microlocale; Sém EDP Ecole Polytechnique; 1988-89.

[GéGo]: P. Gérard, F. Golse: Averaging Results for PDEs under Transversality

Assumptions; in preparation.

[G1]: F. Golse: Remarques sur l'homogénéisation des équations de Transport; C. R. Acad. Sci., 305, (1987), 801-804.

[G2]: F. Golse: Sur les perturbations de systèmes dynamiques et la méthodes de moyennisation en vitesse des EDP; preprint.

[G3]: F. Golse: Quelques résultats de moyennisation pour les équations aux dérivées partielles; Rend. del Semin. Matem. di Torino, Fasc. Speziale 1988.

[G4]: F. Golse: in preparation.

[GLPS] F. Golse, P.-L. Lions, B. Perthame, R. Sentis: Regularity of the Moments of the Solution of a Transport Equation; J. of Funct. Anal., 76, (1988), 110-125.

[GPS] F. Golse, B. Perthame, R. Sentis: Un résultat de compacité pour les équations de transport et application au calcul de la limite de la valeur propre principale d'un opérateur de transport; C.R. Acad. Sci., 301, (1985), 341-344.

[GPo]: F. Golse, F. Poupaud: Limite fluide des équations de Boltzmann des semi-conducteurs pour une statistique de Femi-Dirac; preprint.

[GSe]: F. Golse, R. Sentis: in preparation.

[L]: D. Levermore: Transport in Randomly Mixed Media with Inhomogeneous Anisotropic Statistics; in preparation.

[LPSW]: D. Levermore, G. Pomraning, D. Sanzo, J. Wong: Linear Transport Theory in a Random Medium; J. of Math. Phys., 27, (1986), 2526-2536.

[LPW]: D. Levermore, G. Pomraning, J. Wong: Renewal theory for Transport Processes in Binary Statistical Mixtures; J. of Math. Phys., 29, (1988) 995-1004.

[MT]: F. Murat, L. Tartar:

[T1]: L. Tartar: Compensated Compactness and Applications to Partial Differential Equations; in Nonlinear Analysis and Mechanics, Heriot Watt Symposium, vol IV, R. Knops ed., 136-212, Res. Notes in Math. 39, Pitman, (1979).

[T2]: L. Tartar: Estimations fines de coefficients homogénéisés; De Giorgi Colloquium, P. Kree ed., Pitman.

[Se1]: R. Sentis: Approximation and Homogenization of a Transport Process; SIAM J. of Appl. Math., 39, (1980), 134-141.

[Se2]: R. Sentis: Study of the Corrector of the Principal Eigenvalue of a Transport Operator; SIAM J. of Appl. Math., (1985), 151-166.

[V1]: D. Vanderhaegen: Radiative Transfer in Statistically Heterogeneous Mixtures; J. Quant. Spectrosc. and Radiat. Transfer, 36, (1986), 557.

[V2]: D. Vanderhaegen: Impact of a Mixing Structure on Radiative Transfer in Random Media; J. Quant. Spectrosc. and Radiat. Transfer, 36, (1986), 557;

SOME RECENT RESULTS ON SWIRLING FLOWS OF NEWTONIAN AND NON-NEWTONIAN FLUIDS

K.R. Rajagopal
Department of Mechanical Engineering
University of Pittsburgh
Pittsburgh, PA 15261

1. INTRODUCTION

Karman [1] used a similarity transformation to study the steady axially symmetric flow of a linearly viscous fluid, induced by the rotation of an infinite disk. This work has been followed by extensive studies, which cover a broad spectrum ranging from those concerned with the physics and fluid mechanics of the problem to those which address rigorous mathematical questions regarding existence and uniqueness of solutions. The problems have also been used as test problems for numerical schemes and in the study of matched asymptotic expansions. Such intensive studies notwithstanding, several basic questions regarding the flows remain unanswered and the analysis of the problem is far from complete.

The recent work of Berker [2] and Parter and Rajagopal [3] have revealed that there are other aspects to the problem which have been overlooked. Breaking away from the approach of Karman [1] which assumed axial symmetry, Berker [2] considered the possiblity of solutions that are not necessarily axially symmetric, and established a one-parameter family of solutions for the flow of the classical linearly viscous fluid between two plane parallel disks rotating about a common axis with the <u>same</u> angular speed. The only axially symmetric solution in this family is the rigid body motion; the only solution that would follow from the classical assumptions of Karman. In the light of Berker's work, Parter and Rajagopal [3] re-examined the problem of flow between parallel disks rotating about a common axis with differing angular speeds. Parter and Rajagopal [3] rigorously prove that the axially symmetric solutions are never isolated when considered within the full scope of the Navier-Stokes equation. Similar results apply in the case of the flow due to a single rotating disk and flow due to rotating disks subject to suction or injection at the disk. Based on the existence theorems of Parter and Rajagopal [4], numerical computations have been carried out recently (cf. Lai, Rajagopal and Szeri [4]).

An interesting related problem is the possibility of existence of such asymmetric solutions in the case of viscoelastic fluids. As with the classical linearly viscous fluid, until recently, the investigations have been concerned with the study of axially symmetric solution. Motivated by the work of Berker [2], Rajagopal and Gupta [5, 6, 7] have examined the possibility of the existence of asymmetric solutions for the flow of a special class of viscoelastic fluids between parallel plates rotating with the same angular velocity, about a common axis. More recently, Huilgol and Rajagopal [8] have established the existence of asymmetric solutions for the flow of a popular class of viscoelastic fluids between parallel plates rotating with differing angular velocities.

The results of Berker [2] have relevance to another very interesting application in fluid dynamics, the flow occurring in the orthogonal rheometer (cf. Maxwell and Chartoff [9]). The apparatus consists of two parallel disks which rotate with the same constant angular speed about two parallel but different axes. The fluid to be tested fills the space between the plates. If the fluid is non-Newtonian, then normal stresses develop due to the flow and measuring these will help in characterizing the fluid that fills the apparatus. The motion occurring in such an apparatus has been studied by several authors and all the early works ignored the inertial effects in the treatment of the problem (cf. Huilgol [10]).

Rajagopal [11] recognized that a velocity field, similar to that used by Berker [2], can be employed in this problem and that the velocity field assumed by Berker [2] was a motion with constant principal relative stretch history. He used this fact to show that the flow of any homogeneous incompressible simple fluid in such a configuration is governed by a second order partial differential equation. Thus, unlike other boundary value problems in which one might require additional boundary condition for specific non-Newtonian fluid models (of the differential type), the adherence boundary condition is sufficient for determinacy. The problem being well posed, one can discuss existence, uniqueness and other related questions. Detailed numerical computations have been carried out recently for specific integral constitutive models and (cf. Rajagopal et al. [12], Bower et al. [13]).

2. FLOW BETWEEN PARALLEL DISKS ROTATING ABOUT A COMMON AXIS

We shall first briefly discuss the axially symmetric flow of a Navier-Stokes fluid due to two infinite parallel disks rotating with differing angular speeds about a common axis.

Karman assumed an axially symmetric velocity field of the form (cf. Figure 1):

$$v_r = \frac{r}{2} H'(z) \quad , \quad v_\theta = \frac{r}{2} G(z) \quad \text{and} \quad v_z = -H(z) \quad . \tag{2.1}$$

Here, v_r, v_θ and v_z denote the components of the velocity in the r, θ and z directions, respectively. Notice that the velocity field (2.1) automatically satisfies the constraint of incompressiblity. Substituting (2.1) into the Navier-Stokes equations yields

$$\varepsilon H^{iv} + HH''' + GG' = 0 \quad , \tag{2.2}$$

and

$$\varepsilon G'' + HG' - H'G = 0 \quad , \tag{2.3}$$

where $\varepsilon \equiv \frac{\mu}{\rho}$.

If we are interested in the flow due to the two rotating disks at $z = h$ and $z = 0$, then (2.2) and (2.3) would be valid in the interval $0 \leq z \leq h$. The appropriate boundary conditions for the problem are

$$H(0,\varepsilon) = H(h,\varepsilon) = 0 \quad \text{(no penetration)} \tag{2.4}$$

$$H'(0,\varepsilon) = H'(h,\varepsilon) = 0 \quad \text{(adherence)} \tag{2.5}$$

$$G(0,\varepsilon) = 2\Omega_0 \quad , \quad G(h,\varepsilon) = 2\Omega_{+h} \quad , \tag{2.6}$$

where Ω_0 and Ω_{+h} are the angular speeds of the disks at $z = 0$ and $z = h$, respectively.

Batchelor [14] was the first to study the above boundary value problem. He predicted that at high Reynolds numbers, i.e., small values of ε, boundary layers would develop on each disk with the core of the fluid rotating at a constant angular speed. Stewartson [15], in analysing the same problem, reasoned that at large Reynolds numbers the flow in the core region would be purely axial. Thus, the stage was set for determining whether the solutions predicted by Batchelor [14] or Stewartson [15], or both such solutions, were possible.

Existence and uniqueness of solutions for large ε has been established by Hastings [16] and Elcrat [17]. While Parter and McLeod [18] have proved existence of solution for counter rotating disks for $\varepsilon > 0$, and Kreiss and Parter [19] have shown the existence and multiplicity of solutions for co-rotating disks for $0 < \varepsilon << 1$, the general question of existence remains unproven.

There have been several numerical and asymptotic studies of the above boundary value problem and they are too numerous to detail here. A discussion of the same can be found in the reivew article by Rajagopal [20].

The earliest experiment on the flow between two rotating disks was carried out by Stewartson [15]. He studied the flow of air due to the rotation of two 6 inch diameter cardboard disks to obtain some qualitative information regarding the flow. Experimental investigations have also been carried out by Mellor, Chapple and Stokes [21]. Recently, Szeri, Schneider, Labbe and Kaufman [22] have been able to duplicate the velocity field conjectured by Batchelor [14]. However, they were unable to observe the several other solutions predicted both numerically and asymptotically.

We now turn our attention to the axially symmetric flow of non-Newtonian fluids between parallel disks rotating about a common axis. Phan-Thien [23] studied the time-dependent flow of a Maxwell fluid between two disks rotating about a common axis. Recently, Huilgol and Keller [25] have studied the steady flow of an Oldroyd-B fluid between disks rotating about a common axis, with a view of establishing multiple solutions.

Next, let us consider the case $\Omega_h = \Omega_0 = \Omega \neq 0$, within the context of the Karman equations (2.2) - (2.6). Then, the only solution to equations (2.2) - (2.6) is the rigid body solution:

$$G \equiv 2\Omega \quad \text{and} \quad H \equiv 0, \tag{2.7}$$

which is isolated and stable. By isolated, we mean that there is a neighborhood of this solution wherein there are no other solutions, and by stable we mean there is no bifurcation from this solution, in particular the linearized problem at this solution is non-singular.

Recently, Berker [2] established a truly remarkable result for the problem of two infinite parallel disks rotating with the same angular velocity Ω, about a common axis. He sought solutions of the form

$$v_x = -\Omega[y - f(z)] \quad , \tag{2.8}$$

$$v_y = \Omega[z - g(z)] \quad , \tag{2.9}$$

$$v_z = 0 \quad , \tag{2.10}$$

where v_x, v_y and v_z are the components of the velocity in the x, y and z directions, respectively. Such a velocity field corresponds to a flow wherein streamlines in any z = constant plane are concentric circles, with no flow across any such plane; the locus of the centers of these circles as the z = constant plane shifts from z = 0 to z = h being a curve in space described by x = f(z) and y = g(z). Notice that the velocity field (2.8)-(2.10) automatically satisfies the constraint of incompressiblity.

Substituting (2.8)-(2.10) into the Navier-Stokes equation, and taking the curl of the equation we obtain

$$\mu f''' + \rho\Omega g' = 0 \quad , \tag{2.11}$$

$$\mu g''' - \rho\Omega f' = 0 \quad . \tag{2.12}$$

The appropriate boundary conditions are

$$f(h) = 0, \quad f(0) = 0, \quad g(0) = 0, \quad g(h) = 0. \tag{2.13}$$

Since we have increased the order of the equation by taking the curl, we need to augment the boundary conditons provided above. Let the locus of the centers rotation x = f(z), y = g(z) intersect the mid-plane z = h/2 at $(\ell,0)$, i.e.,

$$f(h/2) = \ell, \qquad g(h/2) = 0 \ . \tag{2.14}$$

Thus, we need to seek a solution to (2.11) and (2.12) subject to (2.13) and (2.14).

When $\ell = 0$, we obtain the rigid body solution. However, this is but a one-parameter family of solutions that are possible for each value of ℓ. Moreover, when $\ell \neq 0$, the solutions are not axially symmetric. Thus, in the special case when $\Omega_h = \Omega_0$, the axially symmetric solution of the Karman equation is imbedded in a much larger class of solutions. This naturally leads us to ask the question whether the axially symmetric solutions to the Karman equations are imbedded in a larger class of solutions when $\Omega_h \neq \Omega_0$? This question has been recently answered by Parter and Rajagopal [3].

Parter and Rajagopal [3] assumed a velocity field of the form

$$v_x = \frac{x}{2} H'(z) - \frac{y}{2} G(z) + g(z) \quad , \tag{2.15}$$

$$v_y = \frac{y}{2} H'(z) + \frac{x}{2} G(z) - f(z) \quad , \tag{2.16}$$

and

$$v_z = -H(z) \quad . \tag{2.17}$$

The above velocity field in cylindrical co-ordinates has the form

$$v_r = \frac{r}{2} H'(z) + g(z) \cos\theta - f(z) \sin\theta \quad , \tag{2.18}$$

$$v_\theta = \frac{r}{2} G(z) - g(z) \sin\theta - f(z) \cos\theta \quad , \tag{2.19}$$

$$v_z = -H(z) \quad . \tag{2.20}$$

Notice that when $f \equiv 0$, $g \equiv 0$, we recover the velocity field assumed by Karman. When $H \equiv 0$ and $G \equiv 2\Omega$, we obtain the velocity field assumed by Berker.

Substituting equations (2.21)-(2.23) into the Navier-Stokes equation yields

$$\varepsilon H^{iv} + HH''' + GG' = 0 \quad , \tag{2.21}$$

$$\varepsilon G'' + HG' - H'G = 0 \quad , \tag{2.22}$$

$$\varepsilon f''' + Hf'' + \frac{1}{2} H'f' - \frac{1}{2} H''f + \frac{1}{2} (Gg)' = 0 \quad , \qquad (2.23)$$

$$\varepsilon g''' + Hg'' + \frac{1}{2} H'g' - \frac{1}{2} H''g - \frac{1}{2} (Gf)' = 0 \quad . \qquad (2.24)$$

The appropriate boundary conditions are

$$H(0,\varepsilon) = H(h,\varepsilon) = 0 \quad , \qquad (2.25)$$

$$H'(0,\varepsilon) = H'(h,\varepsilon) = 0 \quad , \qquad (2.26)$$

$$G(0,\varepsilon) = 2\Omega_0 \quad , \quad G(h,\varepsilon) = 2\Omega_h \quad , \qquad (2.27)$$

$$f(0,\varepsilon) = f(h,\varepsilon) = 0 \quad , \qquad (2.28)$$

$$g(0,\varepsilon) = 0, \; g(h,\varepsilon) = 0 \quad ,. \qquad (2.29)$$

The system of equations (2.24)-(2.27) and boundary conditions (2.28)-(2.32) are underdetermined and as before we can augment the system by requiring

$$f(h/2, \varepsilon) = \ell_1 \; , \; g(h/2, \varepsilon) = \ell_2 \quad . \qquad (2.30)$$

Parter and Rajagopal [3] prove that the axially symmetric solutions are imbedded in a two-parameter family of solutions.

A detailed numerical study of the asymmetric flow between parallel rotating disks has been carried out by Lai, Rajagopal and Szeri [4].

We next turn our attention to discussing solutions that are not axially symmetric in the case of non-Newtonian fluids. In the special case $\Omega_h = \Omega_{-h}$, one-parameter family of solutions that are not axially symmetric have been analytically established by Rajagopal and Gupta [6] in the case of the incompressible fluid of second grade*, by Rajagopal and Wineman [26] in the case of a special subclass of the K-BKZ fluid. Detailed numerical solutions have been carried out in the case of a special subclass of K-BKZ fluids, namely the Wagner and Currie models by Bower, Rajagopal and Wineman [13] and by Rajagopal, Renardy, Renardy and Wineman [12].

When $\Omega_h \neq \Omega_0$, Huilgol and Rajagopal [9] show that a situation similar to that considered by Parter and Rajagopal [3] obtains in the case of an Oldroyd fluid. Huilgol and Rajagopal [9] assume a velocity field of the form (2.18)-(2.20) and show that the problem is governed by four coupled equations for the four functions H, G, f and g.

---

*Earlier, Drouot [26] extended Berker's analysis, and her work clearly implies the possibility of exact solutions in the case of an incompressible homogeneous fluid of second grade. However, she did not solve the specific boundary value problem. In fact Berker's earlier analysis (cf. [29], [30]) implies the exact solution found by Abbott and Walters [27].

These equations are being currently studied with the view of establishing multiplicity of solutions to these equations using analytic continuation methods. Preliminary results do indicate the existence of multiple solutions and limit points.

3. FLOW BETWEEN PARALLEL DISKS ROTATING ABOUT DIFFERENT AXES

We shall first discuss the special case when $\Omega_h = \Omega_0 = \Omega \neq 0$. Let 'a' denote the distance between the parallel but distinct axes (cf. Figure 2). In the case of Navier-Stokes fluid, Abbot and Walters [27] restricted themselves to solutions which possess midplane symmetry and exhibited an exact solution to that problem. However, if we relax the requirement of midplane symmetry, it is easy to show that the problem under consideration possesses a one-parameter family of solutions (cf. Berker [28]). As we mentioned earlier, the above flow has relevance to the flow occurring in an orthogonal rheometer.

We shall now discuss in some detail the flow of an incompressible simple fluid in the orthogonal rheometer. We shall assume that the motion occurring in the orthogonal rheometer has the form (2.8)-(2.10). We shall first show that such a flow is a motion with constant principal relative stretch history. Let $\underset{\sim}{\xi} = (\xi, \eta, \zeta)$ denote the position occupied by a particle $\underset{\sim}{X} = (X, Y, Z)$ at time τ. Let $\underset{\sim}{x} = (x,y,z)$ denote the position occupied by the same particle $\underset{\sim}{X}$ at time t. It follows from (2.8)-(2.10) that

$$\dot{\xi} = -\Omega(\eta - g(\zeta)) \quad , \tag{3.1}$$

$$\dot{\eta} = \Omega(\xi - f(\zeta)) \quad , \tag{3.2}$$

$$\dot{\zeta} = 0 \quad , \tag{3.3}$$

with

$$\xi(t) = x \quad , \quad \eta(t) = y \quad , \quad \text{and} \quad \zeta(t) = z \quad . \tag{3.4}$$

It immediately follows that

$$\xi(\tau) = (x-f(z))\cos\Omega(t-\tau) + (y-g(z))\sin\Omega(t-\tau) + f(z), \tag{3.5}$$

$$\eta(\tau) = -(x-f(z))\sin\Omega(t-\tau) + (y-g(z))\cos\Omega(t-\tau) + g(z), \tag{3.6}$$

$$\zeta(\tau) = z \quad . \tag{3.7}$$

Thus, the history of relative deformation gradient is given by

$$\underset{\sim}{F}_t(t-s) = \begin{matrix} \cos\Omega s & \sin\Omega s & -g'\sin\Omega s + f'(1-\cos\Omega s) \\ -\sin\Omega s & \cos\Omega s & g'(1-\cos\Omega s) + f'\sin\Omega s \\ 0 & 0 & 1 \end{matrix} \quad . \tag{3.8}$$

A motion has constant stretch history if and only if (cf. Noll [31])

$$\underset{\sim}{F}_0(\tau) = \underset{\sim}{Q}(\tau)e^{\tau \underset{\sim}{M}} \quad , \quad \tau \quad (\infty, -\infty) \tag{3.9}$$

$$\underset{\sim}{Q}(0) = \underset{\sim}{1} \tag{3.10}$$

where $\underset{\sim}{Q}(\tau)$ is an orthogonal tensor function and $\underset{\sim}{M}$ is constant in time.

It then follows (cf. Rajagopal [11]) that

$$\underset{\sim}{F}_0(\tau) = e^{\tau \underset{\sim}{L}} \quad , \tag{3.11}$$

where

$$\underset{\sim}{L} \equiv \text{grad } \underset{\sim}{v} = \begin{matrix} 0 & -\Omega & \Omega g' \\ \Omega & 0 & -\Omega f' \\ 0 & 0 & 0 \end{matrix} \quad , \tag{3.12}$$

and hence the flow is one of the constant principal relative stretch history.

The appropriate boundary conditions for the velocity field are

$$u = \frac{\Omega a}{2} - \Omega y \quad , \quad v = \Omega x \quad , \quad w = 0 \qquad \text{at } z = h \quad , \tag{3.13}$$

$$u = -\frac{\Omega a}{2} - \Omega y \quad , \quad v = \Omega x \quad , \quad w = 0 \qquad \text{at } z = 0 \quad , \tag{3.14}$$

and

$$u \rightarrow \mp \infty \quad , \quad v \rightarrow \pm \infty \qquad \text{as } x, y \rightarrow \pm \infty \quad . \tag{3.15}$$

It follows from (3.13), (3.14) and (3.15) that

$$f(h) = f(0) = 0 \quad , \tag{3.16}$$

$$g(h) = \frac{a}{2} \quad , \quad g(0) = -\frac{a}{2} \quad . \tag{3.17}$$

We shall discuss a non-trivial example, namely, the problem associated with the flow of a K-BKZ fluid in an orthogonal rheometer.

The Cauchy stress $\underset{\sim}{T}$ in the K-BKZ fluid has the structure (cf. Kaye [32], Bernstein, Kearsley and Zapas [33]):

$$\underset{\sim}{T} = -p\underset{\sim}{1} + 2\int_{-\infty}^{t} \{U_1 \underset{\sim}{C}_t^{-1}(\tau) - U_2 \underset{\sim}{C}_t(\tau)\} d\tau \quad , \tag{3.18}$$

where

$$\underset{\sim}{C}_t(\tau) = \underset{\sim}{F}_t^T(\tau) \underset{\sim}{F}_t(\tau) \quad . \tag{3.19}$$

In equation (3.18), U denotes the strain energy function for the viscoelastic fluid and is a function of the principal invariants of $\underset{\sim}{C}_t(\tau)$ and $\underset{\sim}{C}_t^{-1}(\tau)$:

$$U = U(I_1, I_2, t - \tau) \tag{3.20}$$

$$I_1 = \text{tr } \underset{\sim}{C}_t(\tau) \quad , \quad I_2 = \text{tr } \underset{\sim}{C}_t^{-1}(\tau) \tag{3.21}$$

and

$$U_i = \frac{\partial U}{\partial I_i} \quad , \quad i = 1, 2 \quad . \tag{3.22}$$

Let

$$s \equiv \text{Sin}\ \Omega(t-\tau) \quad , \quad c \equiv \text{Cos}\ \Omega(t-\tau) \tag{3.23}$$

Then, it follows that

$$I_1(t,\tau) = I_2(t,\tau) = 3 + 2(1-c)(f'^2 + g'^2) \tag{3.24}$$

$$\equiv I(\Omega(t-\tau), z) \quad .$$

The equations of motion reduce to

$$\frac{d}{dz}\{f'B(\kappa) + g'A(\kappa)\} = \rho\Omega^2 f \quad , \tag{3.25}$$

$$\frac{d}{dz}\{-f'A(\kappa) + g'B(\kappa)\} = \rho\Omega^2 g \quad , \tag{3.26}$$

where

$$\kappa \equiv (f'^2 + g'^2)^{1/2} \tag{3.27}$$

and

$$A(\kappa) = 2\int_0^\infty \tilde{U}[3 + 2(1-\cos\Omega\alpha)\kappa^2, \alpha]\sin\Omega\alpha\ d\alpha \quad , \tag{3.28}$$

$$B(\kappa) = 2\int_0^\infty \tilde{U}[3 + 2(1-\cos\Omega\alpha)\kappa^2, \alpha](1-\cos\Omega\alpha)\ d\alpha \quad , \tag{3.29}$$

$$\tilde{U}(I,\alpha) \equiv U_1(I,I,\alpha) + U_2(I,I,\alpha) \quad . \tag{3.30}$$

The above system of equations have been solved for a variety of K-BKZ models by Rajagopal and Wineman [34], Bower et al. [12] and Rajagopal et al. [13]. Figure 3 shows clearly how at sufficiently large Reynolds number boundary layers develop adjacent to the plates, a central region rotating rigidly.

When $\Omega_h \neq \Omega_0$, the flow of a Navier-Stokes fluid between plates rotating about different axes is governed by the same system of differential equations (2.21)-(2.24). The only difference in the boundary value problem from that governing the flow about a common axis occurs in the specification of boundary conditions. The boundary conditions $(2.30)_1$ and $(2.30)_2$ are replaced by

$$g(-h,\varepsilon) = -\frac{a\Omega_0}{2} \quad , \quad g(h,\varepsilon) = \frac{a\Omega_{+h}}{2} \quad . \tag{3.31}$$

Parter and Rajagopal [4] have discussed questions regarding the existence of solutions to the system (2.21)-(2.24), (2.25)-

(2.28), (2.30) and (3.31). Numerical solutions of the above system have been carried out by Lai, Rajagopal and Szeri [13]. In this case, the locus of the stagnation points is far from simple and once again does not possess any mid-plane symmetry (cf. Figure 4).

When $\Omega_0 \neq \Omega_h$, not much work has been done in the case of non-Newtonian fluids. Huilgol and Rajagopal [9] have derived the equations governing the flow for an Oldroyd-B fluid. These equations are being solved numerically using an analytic continuation technique.

ACKNOWLEDGEMENT

The support of the National Science Foundation and AFOSR is gratefully acknowledged.

BIBLIOGRAPHY

[1] T. von Karman, Uber laminare und turbulente Reibung, Z. Angen. Math. Mech. 1, 232-252 (1921).

[2] R. Berker, A new solution of the Navier-Stokes equation for the motion of a fluid contained between two parallel planes rotating about the same axis, Archiwum Mechaniki Stosowanej, 31, 265-280 (1979).

[3] S.V. Parter and K.R. Rajagopal, Swirling flow between rotating plates, Arch. Ratl. Mech. Anal., 86, 305-315 (1984).

[4] C.Y. Lai, K.R. Rajagopal and A.Z. Szeri, Asymmetric flow between parallel rotating disks, J. Fluid Mech., 146, 203-225 (1984).

[5] K.R. Rajagopal and A.S. Gupta, Flow and stability of second grade fluids between two parallel rotating plates, Archiwum Mechaniki Stosowanej, 33, 663-674 (1981).

[6] K.R. Rajagopal and A.S. Gupta, Flow and stability of a second grade fluid between two parallel rotating plates about non-coincident axes, Intl. J. Eng. Science, 19, 1401-1409 (1985).

[7] K.R. Rajagopal and A.S. Wineman, A class of exact solutions for the flow of a viscoelastic fluid, Archiwum Mechaniki Stosowanej, 35 747-752 (1983).

[8] R.R. Huilgol and K.R. Rajagopal, Non-Axisymmetric flow of a viscoelastic fluid between rotating disks, submitted for publication.

[9] B. Maxwell and R.P. Chartoff, Studies of a polymer melt in an orthogonal rheometer, Trans. Soc. Rheology, 9, 51-52 (1965).

[10] R.R. Huilgol, On the properties of the motion with constant stretch history occuring in the Maxwell Rheometers, Trans. Soc. Rheol., 13, 513-526 (1969).

[11] K.R. Rajagopal, On the flow of a simple fluid in an orthogonal rheometer, Arch. Ratl. Mech. Anal., 79, 29-47 (1982).

[12] M. Bower, K.R. Rajagopal and A.S. Wineman, A numerical study of the inertial effects of the flow of a shear thinning K-BKZ fluid between rotating parallel plates, submitted for publication.

[13] K.R. Rajagopal, M. Renardy, Y. Renardy and A.S. Wineman, Flow of viscoelastic fluids between plates rotating about distinct axes, In Press, Rheologica Acta.

[14] G.K. Batchelor, Note on a class of solutions of the Navier-Stokes equations representing steady rotationally-symmetric flow, Quart. J. Mech. Appl. Math., 4, 29-41 (1951).

[15] K. Stewartson, On the flow between two rotating co-axial disks, Proc. Cambridge Philos. Soc., 49, 333-341 (1953).

[16] S.P. Hastings,On existence theorems for some problems from boundary layer theory, Arch. Ratl. Mech. Anal., 38, 308-316.

[17] A.R. Elcrat, On the swirling flow between rotating co-axial disks, J. Differential Equations, 18, 423-430 (1975).

[18] J.B. McLeod and S.V. Parter, On the flow between two counter-rotating infinite plane disks, Arch. Ratl. Mech. Anal., 54, 301-327 (1974).

[19] H.O. Kreiss and S.V. Parter, On the swirling flow between rotating co-axial disks: existence and non-uniqueness, Comm. Pure and Applied Math., 36, 35-84 (1983).

[20] G.L. Mellor, P.J. Chapple and V.K. Stokes, On the flow between a rotating and a stationary disk, J. Fluid Mech., 31, 95-112 (1968).

[21] D. Dijkstra and G.J.F. van Heijst, The flow between finite rotating disks enclosed by a cylinder, J. Fluid Mech., 128, 123-154 (1983).

[22] A.Z. Szeri, S.J. Schneider, F. Labbe and H.N. Kaufmann, Flow between rotating disks. part 1. Basic flow, J. Fluid Mech., 134, 103-131 (1983).

[23] N. Phan-Thien, Co-axial disk flow and flow about a rotating disk of a Maxwellian fluid, J. Fluid Mech., 128, 427-442.

[24] N. Phan-Thien, Co-axial disk flow of an Oldroyd B-Fluid: Exact solution and stability, J. Non-Newt. Fluid Mech., 13, 325-340 (1983).

[25] R.R. Huilgol and J.B. Keller, Flow of viscoelastic fluids between rotating disks: Part 1, J. Non-Newt. Fluid Mech., 18, 110 (1975).

[26] R. Drouot, Sur un cas d'integration des dquations du mouvement d'un fluide incompressible du deuxieme ordre, C.R. Acad. Sc. Paris, 265 A, 300-304 (1967).

[27] T.N.G. Abbot and K. Walters, Rheometrical flow systems, Part 2, Theory for the orthogonal rheometer, including an exact solution fo the Navier-Stokes equations, J. Fluid Mech., 40, 205-213 (1970).

[28] R. Berker, An exact solution of the Navier-Stokes equation, the vortex with curvilinear axis, Intl. J. Eng. Science, 20, 217-230 (1982).

[29] R. Berker, Integration des equations du mouvement d'un fluide visquex, incompressible, Hanbuch der Physik, VIII/2, Berlin-Gottingen-Heidelberg, Springer (1963).

[30] R. Berker, Sur quelques cas d'integration des equations du mouvement d'un fluide incompressible, Lille, Paris (1936).

[31] W. Noll, Motions with constant stretch history, Arch. Rational Mech., Anal., 11, 97-105 (1962).

[32] A. Kaye, Note No. 134, College of Aeronautics Cranfield Institute of Technology (1962).

[33] B. Bernstein, E.A. Kearsley and L.J. Zapas, A study of stress relaxation with finite strain, Trans. Soc. Rheol., 7, 391-410 (1963).

[34] K.R. Rajagopal and A.S. Wineman, Flow of a BKZ fluid in an orthogonal rheometer, Journal of Rheology, 27, 509-516 (1983).

EVAPORATION AND CONDENSATION OF A RAREFIED GAS BETWEEN ITS TWO PARALLEL PLANE CONDENSED PHASES WITH DIFFERENT TEMPERATURES AND NEGATIVE TEMPERATURE-GRADIENT PHENOMENON — NUMERICAL ANALYSIS OF THE BOLTZMANN EQUATION FOR HARD-SPHERE MOLECULES —

Yoshio Sone, Taku Ohwada, and Kazuo Aoki
Department of Aeronautical Engineering
Kyoto University, Kyoto 606, Japan

ABSTRACT

A rarefied gas between its two parallel plane condensed phases is considered, and its steady behavior, especially the rate of evaporation or condensation on the condensed phases and the negative temperature-gradient phenomenon, is studied numerically on the basis of the linearized Boltzmann equation for hard-sphere molecules under the conventional boundary condition and its generalization. The method of analysis is the finite-difference method developed recently by the authors. Not only the temperature and density distributions and the mass and energy fluxes in the gas but also the velocity distribution function of the gas molecules is obtained with good accuracy for the whole range of the Knudsen number.

I. INTRODUCTION

Consider a rarefied gas between its two parallel plane condensed phases with different temperatures. The gas evaporates on the condensed phase with higher temperature and condenses on the other. This simple and fundamental problem of evaporation and condensation draws special attention in connection with the negative temperature-gradient phenomenon.[1-9] The analyses of the problem are, however, based on the Boltzmann-Krook-Welander (BKW) equation or crude assumptions. Thus, its accurate analysis on the basis of the standard Boltzmann equation is required but has not yet been carried out because of the complex collision integral in the Boltzmann equation.

In this paper we try to carry out the accurate analysis of this two-surface problem of evaporation and condensation for the whole range of the Knudsen number on the basis of the linearized Boltzmann equation for hard-sphere molecules. The method used in the analysis is the finite-difference method developed in the temperature-jump problem by the authors,[10] where an efficient way of computation of the linearized collision integral is proposed and its universal numerical data useful for analyzing various problems are stored. We first consider the problem under the conventional boundary condition on the condensed phase specified in Sec. II, for which the velocity distribution function of the molecules leaving the condensed phase is independent of that of the incident molecules. Then, we discuss the problem under a generalized boundary condition suggested by the experiment by Wortberg[9] and show that the solution for the generalized condition is derived from that for the conventional condition by simple formulae.

II. PROBLEM AND ASSUMPTION

Consider a rarefied gas between its two parallel infinite plane condensed phases at rest with different temperatures. Let the temperature of one of the condensed phases [at $X_1 = -D/2$, $(D > 0)$; X_i is the rectangular space coordinate system] be $T_0(1 - \Delta\tau)$ and that of the other (at $X_1 = D/2$) be $T_0(1 + \Delta\tau)$. We investigate the steady behavior of the gas (the velocity distribution function, the temperature and density distributions, and the mass and energy fluxes in the gas) for the whole range of the Knudsen number (the mean free path of the gas molecules divided by the distance between the condensed phases D) under the following assumptions:

(i) The gas molecules are hard spheres of a uniform size and undergo complete elastic collisions between themselves, and the behavior of the gas is described by the Boltzmann equation.

(ii) In Secs. III - V, we consider the problem under the conventional boundary condition on each condensed phase. That is, the gas molecules leaving each condensed phase constitute the corresponding part of the Maxwellian distribution pertaining to the stationary saturated gas at the temperature [$T_0(1 + \Delta\tau)$ or $T_0(1 - \Delta\tau)$] of the condensed phase. In Sec. VI, we discuss the extention of the solution to the problem under a more general boundary condition.

(iii) The difference of the temperatures of the condensed phases is so small ($|\Delta\tau| \ll 1$) that the governing equation and boundary condition can be linearized around a uniform equilibrium state at rest.

We summarize the remaining main notations used in this paper: ρ_0 is the saturation gas density at temperature T_0; $p_0 = R\rho_0 T_0$ (the saturation pressure at T_0); R is the specific gas constant (the Boltzmann constant divided by the mass of a molecule); ℓ_0 is the mean free path of the gas molecules at the saturated equilibrium state at rest with temperature T_0; $Kn = \ell_0/D$ (the Knudsen number); $k = (\sqrt{\pi}/2)Kn$; $x_i = X_i/D$; $(2RT_0)^{1/2}\zeta_i$ is the molecular velocity; $\zeta = (\zeta_i^2)^{1/2}$; $E(\zeta) = \pi^{-3/2}\exp(-\zeta^2)$; $\rho_0(2RT_0)^{-3/2}E(\zeta)(1 + \phi)$ is the velocity distribution function of the gas molecules; $\rho_0(1 + \omega)$ is the density of the gas; $T_0(1 + \tau)$ is the gas temperature; $p_0(1 + P)$ is the gas pressure; $(2RT_0)^{1/2}u_i$ is the flow velocity; $p_0(\delta_{ij} + P_{ij})$ is the stress tensor (δ_{ij} is Kronecker's delta); $p_0(2RT_0)^{1/2}Q_i$ is the heat flow vector; $p_0(2RT_0)^{1/2}H_i$ is the energy flow vector. Further, the saturation gas pressure at temperature $T_0(1 + \Delta\tau)$ is assumed to be given by $p_0(1 + \beta\Delta\tau)$, where β is a positive constant corresponding to the slope of the Clausius-Clapeyron curve at T_0.

III. BASIC EQUATION

The linearized Boltzmann equation for a steady state in the present spatially one-dimensional case is[11]

$$\zeta_1 \frac{\partial\phi}{\partial x_1} = \frac{1}{k}[L_1(\phi) - L_2(\phi) - \nu(\zeta)\phi], \tag{1}$$

$$L_1(\phi) = \frac{1}{\sqrt{2}\pi}\int \frac{1}{|\zeta_i - \xi_i|}\exp\left(-\xi_i^2 + \frac{|\zeta_i \wedge \xi_i|^2}{|\zeta_i - \xi_i|^2}\right)\phi(x_1, \xi_i)d\vec{\xi}, \tag{2a}$$

$$L_2(\phi) = \frac{1}{2\sqrt{2}\pi}\int |\zeta_i - \xi_i| \exp(-\xi_i^2)\phi(x_1, \xi_i)d\vec{\xi}, \tag{2b}$$

$$\nu(\zeta) = \frac{1}{2\sqrt{2}}\left[\exp(-\zeta^2) + \left(2\zeta + \frac{1}{\zeta}\right)\int_0^{\zeta}\exp(-c^2)dc\right], \tag{2c}$$

where ξ_i is the variable of integration corresponding to ζ_i, $\zeta_i \wedge \xi_i$ is the vector product of ζ_i and ξ_i, $d\vec{\xi} = d\xi_1 d\xi_2 d\xi_3$, and the domain of integration in L_1, L_2, and all the following integrals with respect to the molecular velocity (ζ_i or ξ_i) is the whole molecular velocity space unless otherwise stated.

The boundary condition for Eq. (1) on the condensed phases ($x_1 = \pm 1/2$) is given as

$$\phi(1/2, \zeta_i) = \left(\beta + \zeta^2 - \frac{5}{2}\right)\Delta\tau, \qquad (\zeta_1 < 0), \tag{3a}$$

$$\phi(-1/2, \zeta_i) = -\left(\beta + \zeta^2 - \frac{5}{2}\right)\Delta\tau, \qquad (\zeta_1 > 0). \tag{3b}$$

The solution of the boundary-value problem (1), (3a), and (3b) exists uniquely.[12] From the uniqueness theorem and symmetry of Eqs. (1) - (3b), the solution satisfies

$$\phi(x_1, \zeta_1, \zeta_2, \zeta_3) = -\phi(-x_1, -\zeta_1, \zeta_2, \zeta_3), \tag{4}$$

especially the reflection condition at $x_1 = 0$:

$$\phi(0, \xi_1, \xi_2, \xi_3) = -\phi(0, -\xi_1, \xi_2, \xi_3). \tag{5}$$

Further, in the same way as in Ref. 10, we can show that the solution takes the form

$$\phi = \phi(x_1, \xi_1, \xi_r), \qquad \xi_r = (\xi_2^2 + \xi_3^2)^{1/2}. \tag{6}$$

Thus, the number of the independent variables is reduced to three.

The nondimensional macroscopic variables, ω, τ, u_i, etc., are given as the moments of ϕ:

$$\begin{aligned} &\omega = \int \phi E d\vec{\xi}, && u_i = \int \xi_i \phi E d\vec{\xi}, \\ &\tau = \frac{2}{3}\int (\xi^2 - \frac{3}{2})\phi E d\vec{\xi}, && P = \omega + \tau, \\ &P_{ij} = 2\int \xi_i \xi_j \phi E d\vec{\xi}, && Q_i = \int \xi_i \xi^2 \phi E d\vec{\xi} - \frac{5}{2}u_i, \\ &H_i = \int \xi_i \xi^2 \phi E d\vec{\xi} = Q_i + \frac{5}{2}u_i. \end{aligned} \tag{7}$$

From the conservation equations which are derived from Eq. (1) by multiplying it by E, $\xi_1 E$, or $\xi^2 E$ and integrating the result over the whole molecular velocity space, we have

$$u_1 = \text{const}, \quad P_{11} = 0, \quad Q_1 = \text{const}, \quad H_1 = \text{const}, \tag{8}$$

where Eq. (5) is used for the second equation. These equations will be used for the accuracy test of our numerical computation.

The dependence of the solution on the parameters β and $\Delta\tau$ is very simple. That is, if the solutions for two special values of β, say β_1 and β_2, are known, then the solution for an arbitrary β is given by

$$\phi(\beta=\beta) = \frac{\Delta\tau}{\beta_1 - \beta_2}[(\beta - \beta_2)\phi(\beta=\beta_1, \Delta\tau=1) + (\beta_1 - \beta)\phi(\beta=\beta_2, \Delta\tau=1)]. \tag{9}$$

IV. NUMERICAL ANALYSIS

The boundary-value problem (1), (3a), and (5) over the domain ($0 \le x_1 \le 1/2$) is quite similar to that treated in Ref. 10 [cf. Eqs. (20), (21), and (23) there]. Thus, we can readily make use of the method, scheme, and stored data of computation in Ref. 10.

We determine the solution of the boundary-value problem by pursuing the long-time behavior of the solution of the initial and boundary-value problem [Eq. (1) with the additional $\partial\phi/\partial t$ term on the left-hand side, boundary condition (3a) and (5), and an initial condition (e.g., $\phi = 0$)].

The time-dependent problem is solved numerically by a finite-difference method. The finite-difference scheme for the differential operator is a standard implicit one [cf. Eqs. (33), (34a), and (34b) in Ref. 10]. In the collision integral, the velocity distribution function ϕE is expanded in terms of a system of basis functions such as used in the finite element method. The basis functions are chosen in such a way that ϕE is approximated by a sectionally quadratic function of ζ_1 and ζ_r that takes the exact value at the lattice points in (ζ_1, ζ_r) [cf. Appendix A in Ref. 10]. Thus the collision integral is expressed by the product of the collision integral matrix ($D_{jk\ell m}$ in Ref. 10) and the column vector consisting of the values of ϕE at the lattice points. The collision integral matrix is the collision integral of the basis functions ($\Psi_{\ell m}$ in Ref. 10) at the lattice points and thus is a universal matrix. We have built the matrix for three lattice systems of different fineness in Refs. 10 and 13. Thus, we can effectively compute the collision integral not only for the present problem but also for any problem where the molecular velocity dependence of ϕE is ζ_1 and ζ_r only. Further, the present method of computation is very convenient to vectorize the program of computation for a vector computer such as FACOM VP-400E.

In our finite difference computation, the space interval $0 \leq x_1 \leq 1/2$ is divided into 100 sections, uniform for $k \geq 1$, and nonuniform for $k < 1$ with the minimum width 0.0005 ($k = 0.1$) $\sim$ 0.004 ($k = 0.8$) around $x_1 = 1/2$ and the maximum width 0.0183 ($k = 0.1$) $\sim$ 0.00795 ($k = 0.8$) around $x_1 = 0$. The $\zeta_1\zeta_r$ space is limited to the finite region $(-4.429 \leq \zeta_1 \leq 4.429,\ 0 \leq \zeta_r \leq 4)$, and the region is divided into 36 uniform sections for ζ_r and 116 nonuniform sections for ζ_1 with the minimum width 0.0036 around $\zeta_1 = 0$ and the maximum width 0.2182 around $\zeta_1 = \pm 4.429$. Thus, there are 4329 lattice points in $\zeta_1\zeta_r$ space. This lattice system for $\zeta_1\zeta_r$ is finer than that used in Ref. 10 and is called M3 in Ref. 13. The limitation of ζ_1 and ζ_r to the finite region is legitimate in view of the form of Eqs. (1) - (3), (5), and (7), where the interaction terms of different molecular velocities are multiplied by rapidly decaying functions, and it is confirmed by the computation.

V. RESULT OF ANALYSIS

The temperature and density distributions in the gas are shown in Tables I and II and Figs. 1a, b, and c, where ω and $-\tau$ (or $-\omega$ and τ) are drawn so that the pressure P can be easily read as the difference (or distance) of the two curves. For large β, the temperature gradient in the center region of the gas is in the opposite direction to that of the maintained temperature gradient between the two condensed phases. The possibility of the inverse temperature gradient, called negative temperature-gradient phenomenon, was first noted in the limit $k \to 0$ and has been discussed by various authors.[1-8] Its physical mechanism is discussed in Ref. 2; its thermodynamic discussion is given in Ref. 7. In the limit $k \to 0$, the inverse temperature gradient occurs when

$$\begin{aligned} \beta &> 4.6992, \quad \text{(hard sphere)}, \\ &> 4.7723, \quad \text{(BKW)}, \end{aligned} \tag{10}$$

which is derived by the asymptotic theory of the Boltzmann equation for small Knudsen numbers.[2,10,13-16] The β at the onset of the inverse temperature gradient, β_{rT}, for general k is given in Table III. In the free molecular flow ($k = \infty$), the temperature is uniform ($\tau = 0$) irrespective of β.

The gas velocity u_1, heat flow Q_1, and energy flow H_1 versus k are shown for $\beta = 2$ and 12 in Fig. 2 and Table III. The results for arbitrary β are easily obtained by the same linear combination as in Eq. (9). The mass and energy flows are always from the hotter condensed phase to the colder, but a heat flow in the opposite direction is observed for large β. For large β, $-H_1$ takes its minimum at an intermediate Knudsen number, and around $\beta = 2 \sim 3$, so does $-u_1$. The value of β for which the reversal of the direction of the heat flow occurs is also tabulated as β_{rQ} in Table III. It does not depend much on the Knudsen number. The asymptotic results for small k can be obtained with the aid of the asymptotic theory:[2,10,13,16]

$$\begin{aligned} u_1 &= -\frac{c_1\beta + (c_2 + c_3\beta)k}{1 + c_0 k}\Delta\tau, & Q_1 &= -\frac{5}{2}\kappa\frac{(1 + c_4\beta)k}{1 + c_0 k}\Delta\tau, \\ H_1 &= -\frac{5}{2}\frac{c_1\beta + [c_5 + c_6\beta]k}{1 + c_0 k}\Delta\tau, & \beta_{rT} &= \beta_{rQ} = -c_4^{-1}, \end{aligned} \tag{11}$$

$c_0 = 4.3327$, $c_1 = 0.4670$, $c_2 = -1.0224$, $c_3 = 2.2411$, $c_4 = -0.2128$,

$(c_4^{-1} = -4.6992)$, $\kappa = 1.922284$, $c_5 = c_2 + \kappa$, $c_6 = \kappa c_4 + c_3$,

where all the terms of the asymptotic series are taken. They are also shown in Fig. 2 and Table III. The onset of the inverse heat flow generally differs from that of the inverse temperature gradient but agrees with it in the limit $k \to 0$.

The velocity distribution functions ϕE at three points in the gas are shown for $k = 0.1$, 1, and 10 and for $\beta = 2$, 12, and β_{rQ} in Figs. 3 - 5. For a given β, the velocity distribution function $\phi E/\Delta\tau$ for $\zeta_1 < 0$ at $x_1 = 0.5$ takes the same form for any k [cf. Eq. (3a)]. Thus these are good examples to see the effect of molecular collisions. For large k, the variation of ϕE with respect to x_1 is small except around $\zeta_1 = 0$, where the collision effect is localized, and the variation with respect to ζ_1 is very steep around $\zeta_1 = 0$.

VI. GENERAL BOUNDARY CONDITION

So far we discussed the problem under the conventional boundary condition on the condensed phase, where the velocity distribution function of the molecules leaving the condensed phase is determined by the condition of the condensed phase and is independent of the velocity distribution of the molecules incident on the condensed phase. The generalized boundary condition suggested by Wortberg's experiment[9] showing the dependence of the distribution of the leaving molecules on that of the incident molecules on a solid surface is given as follows. The velocity distribution of the leaving molecules is given by the sum of two terms: α_c times of the distribution stated in (ii) of Sec. II and $(1 - \alpha_c)$ times of the diffuse reflection distribution[17], where α_c $(0 \le \alpha_c \le 1)$ is a constant called condensation factor. In the present problem, from Eqs. (3a) and (3b) and Eq. (3) in Ref. 18, it is given by

$$\phi = (\bar{\beta}_+ + \zeta^2 - \frac{5}{2})\Delta\tau, \qquad (\zeta_1 < 0,\ x_1 = \frac{1}{2}), \tag{12a}$$

$$\bar{\beta}_+ = \alpha_c\beta + (1 - \alpha_c)[\frac{1}{2} + 2\sqrt{\pi}(\Delta\tau)^{-1}\int_{\zeta_1>0}\zeta_1\phi(\frac{1}{2},\ \zeta_i)Ed\boldsymbol{\zeta}],$$

$$\phi = -(\bar{\beta}_- + \zeta^2 - \frac{5}{2})\Delta\tau, \qquad (\zeta_1 > 0,\ x_1 = -\frac{1}{2}), \tag{12b}$$

$$\bar{\beta}_- = \alpha_c\beta + (1 - \alpha_c)[\frac{1}{2} + 2\sqrt{\pi}(\Delta\tau)^{-1}\int_{\zeta_1<0}\zeta_1\phi(-\frac{1}{2},\ \zeta_i)Ed\boldsymbol{\zeta}].$$

The conventional condition corresponds to $\alpha_c = 1$, and the diffuse reflection boundary condition without evaporation and condensation corresponds to $\alpha_c = 0$. The latter problem was treated in Ref. 18.

Under the assumption that α_c takes a common value on both the condensed phases, the solution of Eqs. (1), (12a), and (12b) is easily seen to have the symmetry of Eqs. (4) and (5). Thus

$$\bar{\beta}_+ = \bar{\beta}_- = \bar{\beta}. \tag{13}$$

Noting the symmetry and comparing Eqs. (12a) and (12b) with Eqs. (3a) and (3b), we find that the solution under the generalized boundary condition is expressed by the solution under the conventional condition with different β. That is, let $\phi(x_1,\ \zeta_i;\ \beta)$ be the solution under Eqs. (3a) and (3b), and $\phi_\alpha(x_1,\ \zeta_i;\ \beta_\alpha)$ the solution under Eqs. (12a) and (12b) with a different β (i.e., $\beta = \beta_\alpha$) and the same k. Then the relation between $(\beta,\ \phi)$ and $(\beta_\alpha,\ \phi_\alpha)$ is given by

$$\beta_\alpha = \frac{\beta}{\alpha_c} - \frac{1-\alpha_c}{\alpha_c}[\tfrac{1}{2} + 2\sqrt{\pi}(\Delta\tau)^{-1}\int_{\zeta_1>0}\zeta_1\phi(\tfrac{1}{2},\ \zeta_1;\ \beta)Ed\vec{\zeta}], \qquad (14)$$

$$\phi_\alpha(x_1,\ \zeta_1;\ \beta_\alpha) = \phi(x_1,\ \zeta_1;\ \beta). \qquad (15)$$

In view of Eq. (9), the quantity in the square brackets of Eq. (14) is a linear functions of β. That is,

$$[*] = a(k)\beta + b(k), \qquad (16)$$

where a(k) and b(k) are given in Table III. The β_α increases as α_c decreases when $\beta > b(k)/[1 - a(k)]$. In Table III, $b(k)/[1 - a(k)] < 1$. For β_α given by Eq. (14), the same state as that under the conventional boundary condition is realized under the generalized boundary condition. Thus the inverse temperature gradient or the inverse heat flow is also found under the generalized boundary condition.

Making use of Eq. (9) to eliminate ϕ and β from Eqs. (14) and (15), we have the solution for arbitrary α_c ($\neq 0$) and β_α in terms of two solutions with different β, β_1 and β_2, under the conventional condition as

$$\phi_\alpha(\beta = \beta_\alpha) = \frac{1}{\beta_1-\beta_2}[(A\beta_\alpha + B - \beta_2)\phi(\beta = \beta_1) - (A\beta_\alpha + B - \beta_1)\phi(\beta = \beta_2)], \qquad (17)$$

$$A = \frac{\alpha_c}{1 - a(k)(1 - \alpha_c)}, \qquad B = \frac{b(k)(1 - \alpha_c)}{1 - a(k)(1 - \alpha_c)}. \qquad (18)$$

We can, therefore, obtain the necessary data for an arbitrary α_c from those given in Sec. V.

From the point of solving the problem, the solution for all the cases including $\alpha_c = 0$ is obtained by a linear combination of the solutions of two boundary value problems, say Eqs. (1), (5), and (19a) below and Eqs. (1), (5), and (19b) below.

$$\phi = 1, \qquad (\zeta_1 < 0,\ x_1 = \tfrac{1}{2}), \qquad (19a)$$

$$\phi = \zeta^2, \qquad (\zeta_1 < 0,\ x_1 = \tfrac{1}{2}). \qquad (19b)$$

In the present work, we obtain the data from the solutions say, Φ_1 and Φ_2 of the two boundary-value problems, Eqs. (1), (5), and (19a) and Eqs. (1), (5), and (19b). That is,

$$\phi_\alpha(\beta = \beta_\alpha)(\Delta\tau)^{-1} = (A\beta_\alpha + B - \tfrac{5}{2})\Phi_1 + \Phi_2. \qquad (20)$$

For smaller k the numerical error in the collision integral is more important to the solution. This effect for small k can be reduced by solving the problem in two steps: i) solve Φ_J (J = 1 or 2) and compute corresponding u_1. ii) consider $\tilde{\Phi}_J$:

$$\tilde{\Phi}_J = \Phi_J(\text{new}) - 2\xi_1 u_1(\text{old}), \tag{21}$$

where Φ_J(new) is the final solution Φ_J to be solved and u_1(old) is the preliminary u_1 solved in the first step; derive the equation and boundary condition for $\tilde{\Phi}_J$; and then solve $\tilde{\Phi}_J$, from which Φ_J(new) is obtained.

As the accuracy test of our numerical computation, we examine the constancy of u_1, P_{11}, and H_1 [Eq. (8)] for Φ_1 and Φ_2. The results are:

$$\begin{gathered} V(u_1) < 3.6 \times 10^{-5}, \quad |P_{11}| < 5.4 \times 10^{-5}, \quad V(H_1) < 2.2 \times 10^{-4}, \\ V(u_1) = \max(u_1(x_1)) - \min(u_1(x_1)), \qquad (\text{for each } \Phi_J \text{ and } k). \end{gathered} \tag{22}$$

The computation was carried out by FACOM VP-400E at the Data Processing Center, Kyoto University.

REFERENCES

1. Y. P. Pao, Phys. Fluids **14**, 306 (1971).
2. Y. Sone and Y. Onishi, J. Phys. Soc. Jpn. **44**, 1981 (1978).
3. L. D. Koffman, M. S. Plesset, and L. Lees, Phys. Fluids **27**, 876 (1984).
4. T. Matsushita, Phys. Fluids **19**, 1712 (1976).
5. P. N. Shankar and M. D. Deshpande, Pramāṇa -J. Phys. **31**, L337 (1988).
6. T. Ytrehus and T. Aukrust, in *Rarefied Gas Dynamics*, edited by V. Boffi and C. Cercignani (Teubner, Stuttgart, 1986), Vol. 2, p. 271.
7. L. J. F. Hermans and J. J. M. Beenakker, Phys. Fluids **29**, 4231 (1986).
8. K. Aoki and C. Cercignani, Phys. Fluids **26**, 1163 (1983).
9. R. Mager, G. Adomeit, and G. Wortberg, in *Rarefied Gas Dynamics*, edited by D. P. Weaver, E. P. Muntz, and D. H. Campbell (AIAA, New York, 1989)(to be published).
10. Y. Sone, T. Ohwada, and K. Aoki, Phys. Fluids A **1**, 363 (1989).
11. H. Grad, in *Rarefied Gas Dynamics*, edited by J. A. Laurmann (Academic, New York, 1963), p. 26.
12. C. Cercignani, J. Math. Phys. **8**, 1653 (1967).
13. Y. Sone, T. Ohwada, and K. Aoki, Phys. Fluids A **1** (1989)(to be published).
14. Y. Sone, in *Rarefied Gas Dynamics*, edited by L. Trilling and H. Y. Wachman (Academic, New York, 1969), p. 243.
15. Y. Sone, in *Rarefied Gas Dynamics*, edited by D. Dini (Editrice Tecnico Scientifica, Pisa, 1971), p. 737.
16. Y. Sone and K. Aoki, Transp. Theory Stat. Phys. **16**, 189 (1987); Mem. Fac. Eng. Kyoto Univ. **49**, 237 (1987).
17. C. Cercignani, *The Boltzmann Equation and Its Applications* (Springer, Berlin, 1987).
18. T. Ohwada, K. Aoki, and Y. Sone, in *Rarefied Gas Dynamics*, edited by D. P. Weaver, E. P. Muntz, and D. H. Campbell (AIAA, New York, 1989) (to be published).

Table I. Temperature and density distributions (β=2).

x_1	$\tau/\Delta\tau$							$\omega/\Delta\tau$						
	k=10	k=8	k=6	k=4	k=3	k=2	k=1	k=10	k=8	k=6	k=4	k=3	k=2	k=1
0.0000	0.0000	0.0000	0.0000	0.0000	0.0000	0.0000	0.0000	0.0000	0.0000	0.0000	0.0000	0.0000	0.0000	0.0000
0.0500	0.0072	0.0082	0.0097	0.0120	0.0138	0.0166	0.0219	0.0038	0.0041	0.0044	0.0045	0.0043	0.0035	-0.0003
0.1000	0.0144	0.0165	0.0194	0.0240	0.0277	0.0333	0.0440	0.0076	0.0082	0.0088	0.0091	0.0087	0.0071	-0.0005
0.1500	0.0217	0.0248	0.0292	0.0362	0.0417	0.0502	0.0663	0.0115	0.0124	0.0133	0.0138	0.0133	0.0109	-0.0003
0.2000	0.0291	0.0332	0.0391	0.0485	0.0560	0.0675	0.0892	0.0155	0.0167	0.0179	0.0187	0.0182	0.0151	0.0006
0.2500	0.0367	0.0418	0.0493	0.0612	0.0706	0.0851	0.1126	0.0197	0.0212	0.0228	0.0239	0.0234	0.0199	0.0024
0.3000	0.0444	0.0507	0.0597	0.0742	0.0857	0.1035	0.1371	0.0240	0.0259	0.0280	0.0296	0.0292	0.0254	0.0055
0.3500	0.0524	0.0599	0.0706	0.0879	0.1016	0.1228	0.1630	0.0286	0.0310	0.0336	0.0359	0.0358	0.0321	0.0105
0.4000	0.0609	0.0696	0.0822	0.1024	0.1185	0.1435	0.1911	0.0336	0.0366	0.0399	0.0431	0.0436	0.0405	0.0186
0.4500	0.0700	0.0802	0.0948	0.1184	0.1374	0.1668	0.2232	0.0394	0.0430	0.0474	0.0520	0.0536	0.0519	0.0323
0.5000	0.0813	0.0933	0.1107	0.1391	0.1620	0.1980	0.2680	0.0473	0.0521	0.0583	0.0660	0.0701	0.0728	0.0641

x_1	$\tau/\Delta\tau$ +							$\omega/\Delta\tau$ +						
	k=0.8	k=0.6	k=0.4	k=0.3	k=0.2	k=0.1	k=0.1*	k=0.8	k=0.6	k=0.4	k=0.3	k=0.2	k=0.1	k=0.1*
0.0000	0.0000	0.0000	0.0000	0.0000	0.0000	0.0000	0.0000	0.0000	0.0000	0.0000	0.0000	0.0000	0.0000	0.0000
0.0500	0.0237	0.0261	0.0293	0.0316	0.0349	0.0408	0.0409	-0.0024	-0.0059	-0.0121	-0.0175	-0.0257	-0.0387	-0.0387
0.1000	0.0476	0.0523	0.0589	0.0635	0.0700	0.0818	0.0818	-0.0046	-0.0115	-0.0239	-0.0346	-0.0510	-0.0771	-0.0771
0.1500	0.0718	0.0789	0.0888	0.0958	0.1055	0.1231	0.1230	-0.0064	-0.0165	-0.0351	-0.0509	-0.0756	-0.1150	-0.1150
0.2000	0.0965	0.1061	0.1194	0.1288	0.1417	0.1648	0.1649	-0.0074	-0.0207	-0.0451	-0.0660	-0.0988	-0.1519	-0.1519
0.2500	0.1220	0.1341	0.1510	0.1628	0.1791	0.2074	0.2074	-0.0074	-0.0236	-0.0535	-0.0793	-0.1199	-0.1871	-0.1871
0.3000	0.1485	0.1633	0.1841	0.1984	0.2181	0.2513	0.2514	-0.0058	-0.0245	-0.0594	-0.0896	-0.1377	-0.2194	-0.2194
0.3500	0.1767	0.1944	0.2192	0.2364	0.2596	0.2976	0.2976	-0.0019	-0.0227	-0.0616	-0.0956	-0.1503	-0.2464	-0.2464
0.4000	0.2073	0.2284	0.2578	0.2780	0.3053	0.3480	0.3481	0.0057	-0.0162	-0.0576	-0.0943	-0.1539	-0.2626	-0.2626
0.4500	0.2424	0.2675	0.3025	0.3267	0.3589	0.4074	0.4075	0.0199	-0.0014	-0.0421	-0.0786	-0.1387	-0.2532	-0.2533
0.4625	0.2524	0.2787	0.3155	0.3409	0.3747	0.4252	0.4252	0.0253	0.0048	-0.0348	-0.0703	-0.1290	-0.2419	-0.2419
0.4750	0.2634	0.2910	0.3298	0.3565	0.3922	0.4452	0.4452	0.0321	0.0127	-0.0250	-0.0588	-0.1147	-0.2229	-0.2230
0.4875	0.2758	0.3051	0.3463	0.3747	0.4128	0.4691	0.4692	0.0413	0.0236	-0.0108	-0.0415	-0.0923	-0.1898	-0.1899
0.5000	0.2922	0.3241	0.3692	0.4006	0.4428	0.5064	0.5064	0.0566	0.0430	0.0167	-0.0065	-0.0428	-0.1044	-0.1045

+ The data for $k \leq 0.8$ are interpolated with sufficient accuracy from those at the lattice points, which are not common for different k.

* The data obtained by the asymptotic theory[10,13,16]

Table II. Temperature and density distributions (β=12).

x_1	$\tau/\Delta\tau$							$\omega/\Delta\tau$						
	k=10	k=8	k=6	k=4	k=3	k=2	k=1	k=10	k=8	k=6	k=4	k=3	k=2	k=1
0.0000	0.0000	0.0000	0.0000	0.0000	0.0000	0.0000	0.0000	0.0000	0.0000	0.0000	0.0000	0.0000	0.0000	0.0000
0.0500	-0.0098	-0.0112	-0.0131	-0.0164	-0.0192	-0.0238	-0.0348	0.0778	0.0880	0.1003	0.1180	0.1303	0.1459	0.1634
0.1000	-0.0197	-0.0224	-0.0263	-0.0329	-0.0384	-0.0477	-0.0697	0.1581	0.1764	0.2012	0.2369	0.2616	0.2931	0.3288
0.1500	-0.0297	-0.0338	-0.0397	-0.0495	-0.0578	-0.0718	-0.1048	0.2382	0.2660	0.3034	0.3575	0.3951	0.4433	0.4986
0.2000	-0.0398	-0.0452	-0.0532	-0.0664	-0.0774	-0.0962	-0.1401	0.3197	0.3572	0.4077	0.4811	0.5321	0.5980	0.6755
0.2500	-0.0501	-0.0570	-0.0669	-0.0835	-0.0975	-0.1210	-0.1757	0.4033	0.4508	0.5151	0.6088	0.6744	0.7597	0.8628
0.3000	-0.0607	-0.0690	-0.0811	-0.1012	-0.1181	-0.1464	-0.2120	0.4896	0.5478	0.6269	0.7425	0.8239	0.9311	1.0653
0.3500	-0.0717	-0.0815	-0.0958	-0.1196	-0.1395	-0.1728	-0.2492	0.5801	0.6498	0.7449	0.8847	0.9842	1.1167	1.2900
0.4000	-0.0833	-0.0948	-0.1114	-0.1391	-0.1622	-0.2007	-0.2880	0.6768	0.7592	0.8722	1.0400	1.1607	1.3243	1.5496
0.4500	-0.0962	-0.1095	-0.1287	-0.1606	-0.1872	-0.2313	-0.3299	0.7845	0.8818	1.0160	1.2178	1.3654	1.5703	1.8723
0.5000	-0.1118	-0.1275	-0.1505	-0.1884	-0.2199	-0.2716	-0.3843	0.9217	1.0410	1.2077	1.4641	1.6574	1.9380	2.4056

x_1	$\tau/\Delta\tau$ +							$\omega/\Delta\tau$ +						
	k=0.8	k=0.6	k=0.4	k=0.3	k=0.2	k=0.1	k=0.1*	k=0.8	k=0.6	k=0.4	k=0.3	k=0.2	k=0.1	k=0.1*
0.0000	0.0000	0.0000	0.0000	0.0000	0.0000	0.0000	0.0000	0.0000	0.0000	0.0000	0.0000	0.0000	0.0000	0.0000
0.0500	-0.0395	-0.0465	-0.0586	-0.0688	-0.0847	-0.1108	-0.1108	0.1654	0.1649	0.1583	0.1502	0.1370	0.1231	0.1232
0.1000	-0.0790	-0.0930	-0.1171	-0.1374	-0.1692	-0.2215	-0.2214	0.3332	0.3325	0.3198	0.3038	0.2771	0.2483	0.2484
0.1500	-0.1186	-0.1395	-0.1755	-0.2057	-0.2531	-0.3319	-0.3317	0.5058	0.5057	0.4877	0.4642	0.4242	0.3778	0.3780
0.2000	-0.1585	-0.1862	-0.2337	-0.2736	-0.3363	-0.4415	-0.4413	0.6864	0.6880	0.6661	0.6359	0.5827	0.5152	0.5154
0.2500	-0.1986	-0.2329	-0.2917	-0.3408	-0.4182	-0.5498	-0.5496	0.8787	0.8838	0.8605	0.8248	0.7590	0.6661	0.6663
0.3000	-0.2393	-0.2800	-0.3493	-0.4072	-0.4983	-0.6558	-0.6555	1.0882	1.0994	1.0784	1.0396	0.9629	0.8398	0.8400
0.3500	-0.2807	-0.3275	-0.4067	-0.4724	-0.5758	-0.7575	-0.7572	1.3230	1.3447	1.3323	1.2944	1.2109	1.0545	1.0548
0.4000	-0.3236	-0.3760	-0.4638	-0.5362	-0.6494	-0.8510	-0.8506	1.5979	1.6373	1.6445	1.6155	1.5351	1.3500	1.3501
0.4500	-0.3694	-0.4269	-0.5216	-0.5982	-0.7167	-0.9273	-0.9269	1.9459	2.0176	2.0682	2.0666	2.0158	1.8365	1.8362
0.4625	-0.3818	-0.4404	-0.5364	-0.6136	-0.7320	-0.9412	-0.9407	2.0538	2.1381	2.2072	2.2188	2.1853	2.0248	2.0245
0.4750	-0.3949	-0.4547	-0.5517	-0.6290	-0.7465	-0.9514	-0.9509	2.1769	2.2772	2.3708	2.4008	2.3934	2.2694	2.2689
0.4875	-0.4095	-0.4702	-0.5680	-0.6448	-0.7600	-0.9558	-0.9552	2.3258	2.4476	2.5763	2.6343	2.6695	2.6188	2.6182
0.5000	-0.4281	-0.4903	-0.5883	-0.6637	-0.7731	-0.9458	-0.9451	2.5440	2.7087	2.9104	3.0310	3.1723	3.3563	3.3555

+ The data for $k \leq 0.8$ are interpolated with sufficient accuracy from those at the lattice points, which are not common for different k.

* The data obtained by the asymptotic theory[10,13,16]

Table III. u_1, Q_1, H_1, β_{rT}, β_{rQ}, a(k), and b(k).

k	$\beta = 2$			$\beta = 12$						
	$u_1/\Delta\tau$	$Q_1/\Delta\tau$	$H_1/\Delta\tau$	$u_1/\Delta\tau$	$Q_1/\Delta\tau$	$H_1/\Delta\tau$	β_{rT}	β_{rQ}	a(k)	b(k)
∞	-0.8463	-0.7052	-2.821	-6.488	2.116	-14.10	——	4.5	-1	1
20.00	-0.8428	-0.6908	-2.798	-6.425	2.046	-14.02	6.185	4.524	-0.9789	0.9702
15.00	-0.8419	-0.6864	-2.791	-6.408	2.026	-13.99	6.204	4.530	-0.9730	0.9616
10.00	-0.8404	-0.6779	-2.779	-6.376	1.990	-13.95	6.227	4.541	-0.9625	0.9456
8.00	-0.8395	-0.6719	-2.771	-6.355	1.965	-13.92	6.235	4.549	-0.9553	0.9346
6.00	-0.8383	-0.6624	-2.758	-6.324	1.926	-13.88	6.237	4.559	-0.9446	0.9175
4.00	-0.8366	-0.6448	-2.736	-6.271	1.858	-13.82	6.219	4.576	-0.9264	0.8872
3.00	-0.8356	-0.6288	-2.718	-6.227	1.799	-13.77	6.187	4.590	-0.9113	0.8606
2.00	-0.8348	-0.6003	-2.687	-6.158	1.700	-13.70	6.109	4.610	-0.8872	0.8152
1.50	-0.8350	-0.5753	-2.663	-6.106	1.617	-13.65	6.025	4.625	-0.8684	0.7770
1.00	-0.8369	-0.5328	-2.625	-6.029	1.482	-13.59	5.863	4.644	-0.8406	0.7143
0.80	-0.8391	-0.5058	-2.603	-5.987	1.400	-13.57	5.752	4.654	-0.8250	0.6755
0.60	-0.8432	-0.4672	-2.575	-5.936	1.286	-13.55	5.590	4.665	-0.8052	0.6214
0.40	-0.8516	-0.4069	-2.536	-5.869	1.113	-13.56	5.334	4.678	-0.7787	0.5387
0.30	-0.8592	-0.3613	-2.509	-5.828	0.9844	-13.59	5.148	4.685	-0.7615	0.4771
0.20	-0.8716 (-0.8711	-0.2960 -0.2958	-2.475 -2.474)*	-5.780 (-5.774	0.8034 0.8000	-13.65 -13.64)	4.914 (4.6992	4.692 4.6992)	-0.7399 (-0.7382	0.3898 0.3884)
0.15	-0.8808 (-0.8806	-0.2510 -0.2510	-2.453 -2.453)	-5.751 (-5.749	0.6800 0.6788	-13.70 -13.69)	4.788 (4.6992	4.696 4.6992)	-0.7263 (-0.7257	0.3301 0.3295)
0.10	-0.8931 (-0.8930	-0.1926 -0.1926	-2.425 -2.425)	-5.716 (-5.715	0.5211 0.5209	-13.77 -13.77)	4.692 (4.6992	4.698 4.6992)	-0.7095 (-0.7094	0.2531 0.2529)
0.08	(-0.8991	-0.1640	-2.412)	(-5.698	0.4436	-13.80)	(4.6992	4.6992)	(-0.7014	0.2153)
0.06	(-0.9060	-0.1314	-2.397)	(-5.678	0.3555	-13.84)	(4.6992	4.6992)	(-0.6923	0.1726)
0.04	(-0.9140	-0.0941	-2.379)	(-5.658	0.2545	-13.89)	(4.6992	4.6992)	(-0.6818	0.1236)
0.02	(-0.9232	-0.0508	-2.359)	(-5.633	0.1374	-13.95)	(4.6992	4.6992)	(-0.6698	0.0667)
0.01	(-0.9284	-0.0265	-2.347)	(-5.619	0.0716	-13.98)	(4.6992	4.6992)	(-0.6629	0.0347)
0.00	(-0.9340	0.0000	-2.335)	(-5.604	0.0000	-14.01)	(4.6992	4.6992)	(-0.6556	0.0000)

* The data in the parentheses are obtained by the asymptotic theory[10,13,16] [cf. Eq. (11)].

(a) $\beta = 2$

(b) $\beta = 12$

(c) $\beta = \beta_{rT}$:

$$\beta_{rT} = \begin{cases} 4.692 \ (k=0.1) \\ 5.863 \ (k=1) \\ 6.227 \ (k=10) \end{cases}$$

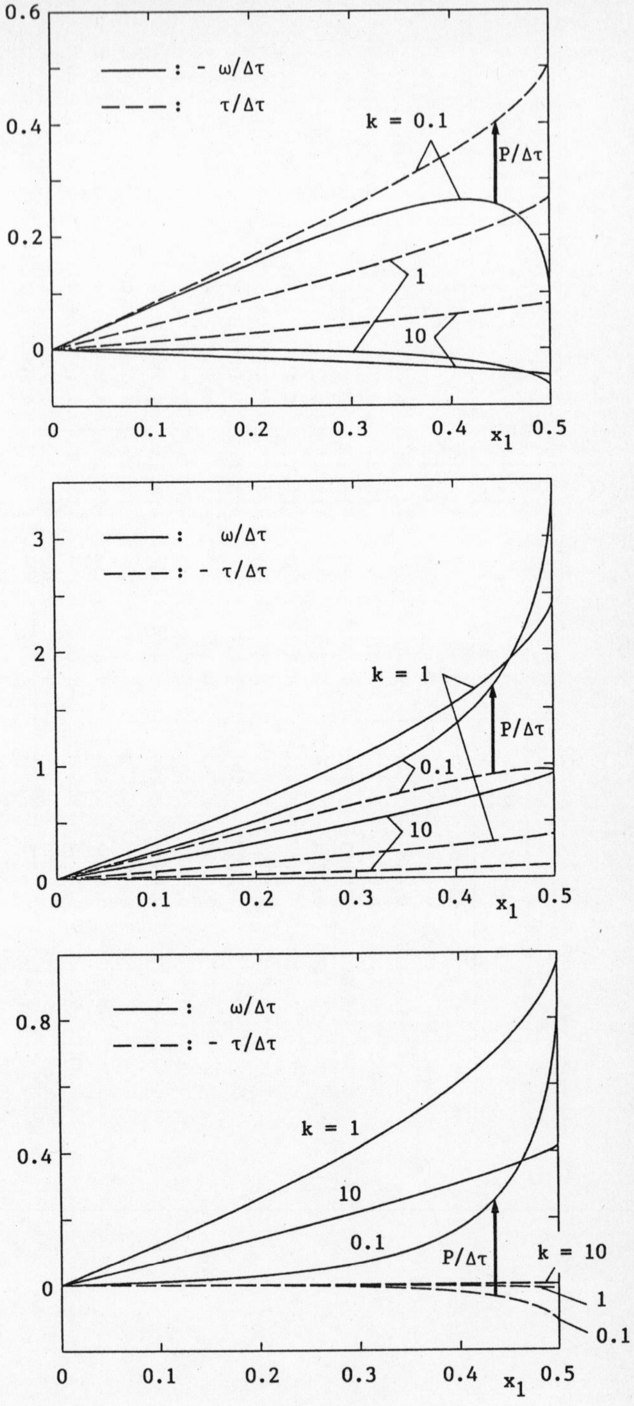

Fig. 1. Temperature and density distributions.

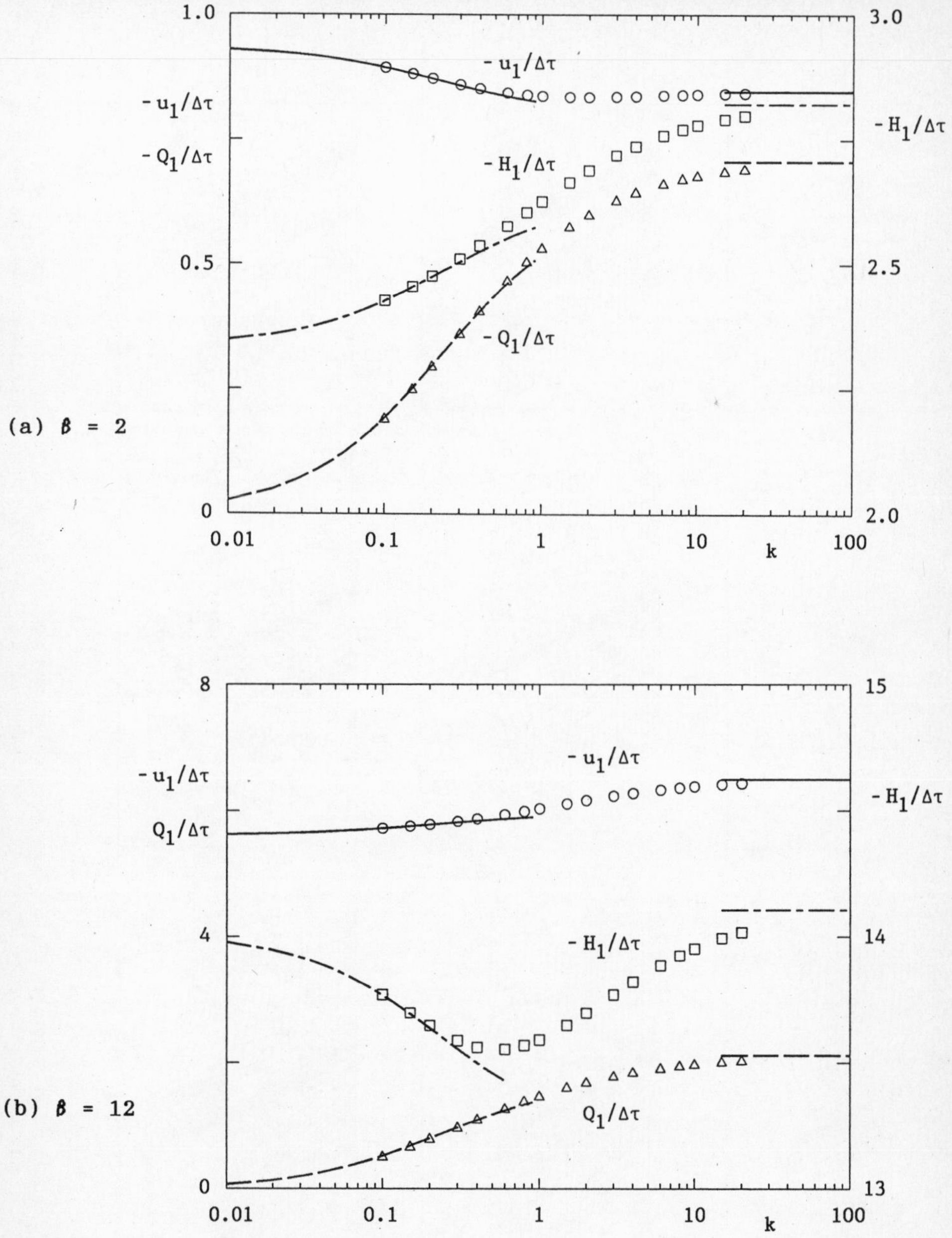

Fig. 2. u_1, Q_1, and H_1. ○, △, and □ indicate the numerical results. The curves for small k indicate the analytical results [Eq. (11)] by the asymptotic theory[10,13,16]. The straight lines for large k indicate the results for the free molecular flow.

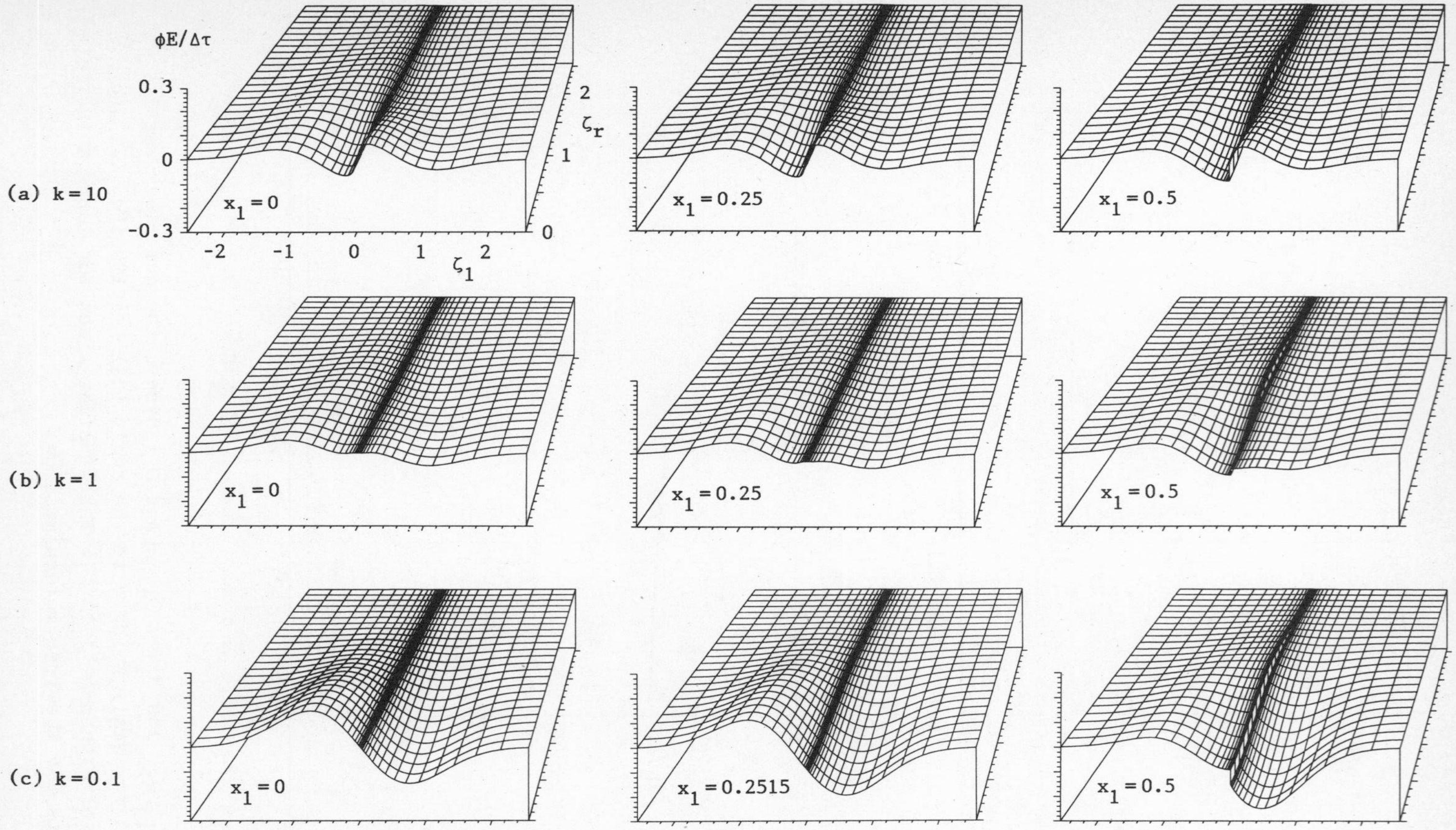

Fig. 3. The velocity distribution function ϕE $(\beta = 2)$. The scales are common for all the figures. Each three $\zeta_1 = \text{const}$ and each $\zeta_r = \text{const}$ lattices in our computation are shown on the surface $\phi E/\Delta\tau$. The $\phi E/\Delta\tau$ is negligibly small at the four corners in the figures, and therefore they may be taken as the reference points.

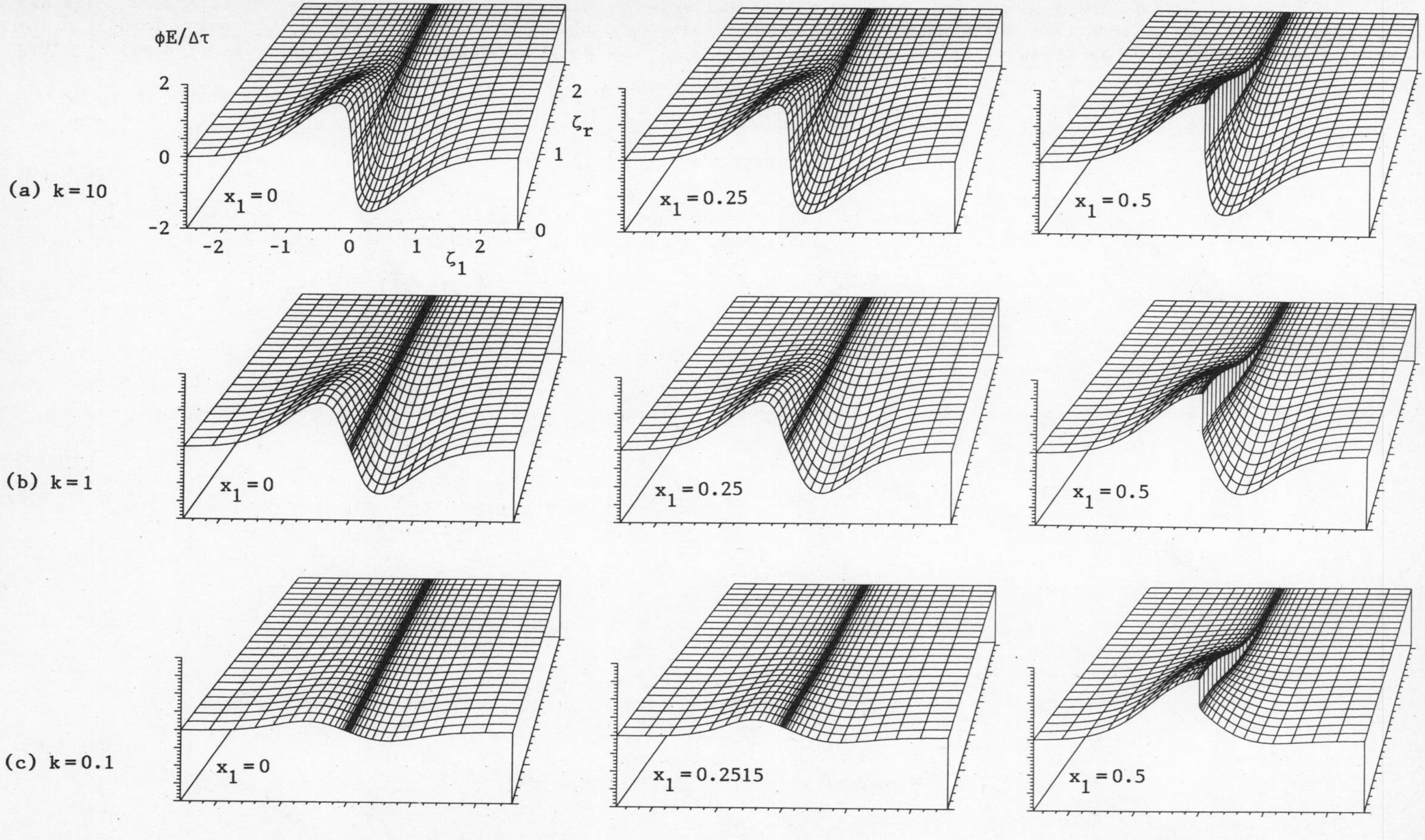

Fig. 4. The velocity distribution function ϕE ($\beta = 12$). The scales are common for all the figures. Each three $\zeta_1 = \text{const}$ and each $\zeta_r = \text{const}$ lattices in our computation are shown on the surface $\phi E/\Delta\tau$. The $\phi E/\Delta\tau$ is negligibly small at the four corners in the figures, and therefore they may be taken as the reference points.

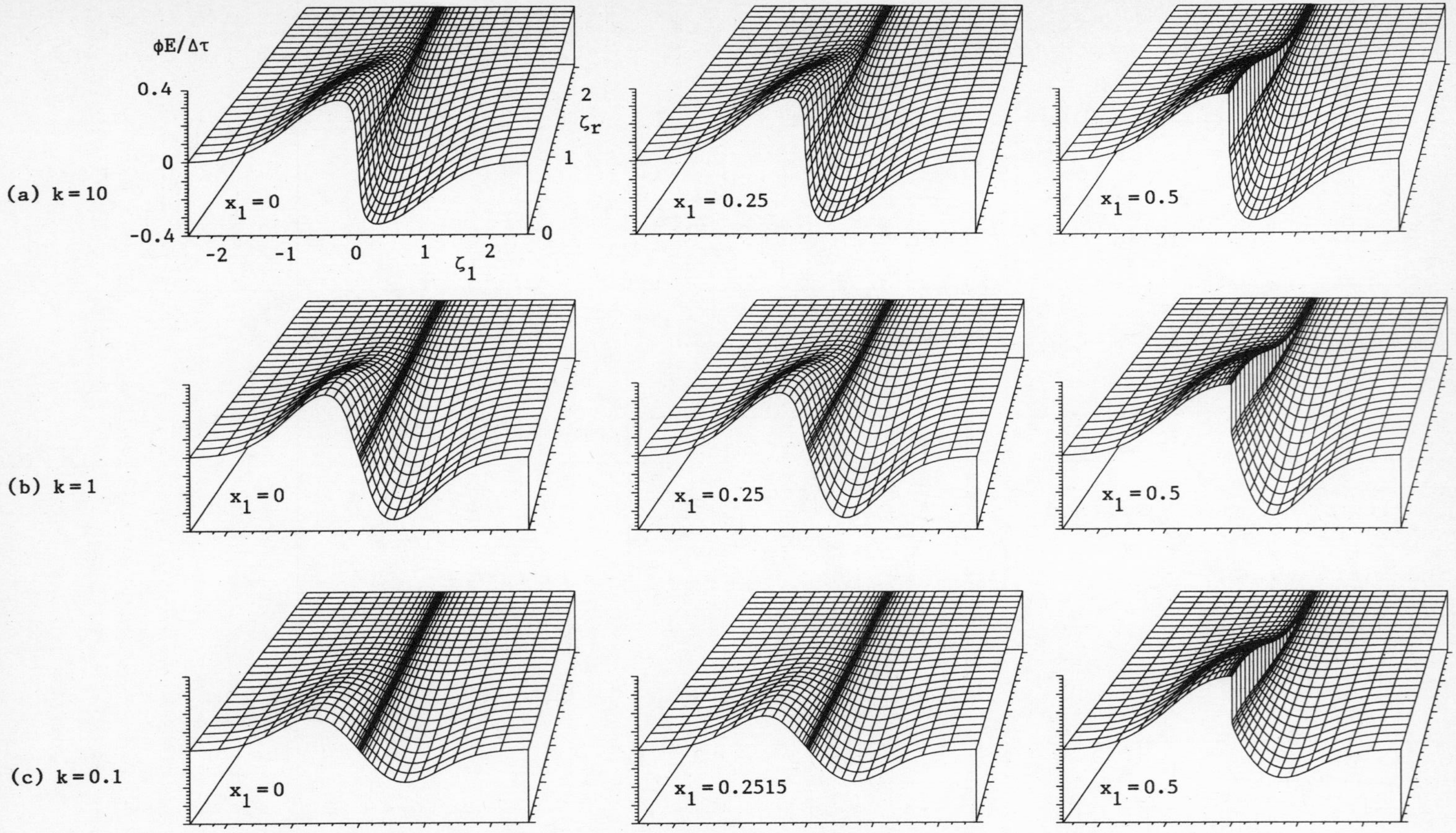

Fig. 5. The velocity distribution function ϕE [$\beta = \beta_{rQ}$: $\beta_{rQ} = 4.541$ (k = 10), 4.644 (k = 1), 4.698 (k = 0.1)]. The scales are common for all the figures. Each three $\zeta_1 = \text{const}$ and each $\zeta_r = \text{const}$ lattices in our computation are shown on the surface $\phi E/\Delta\tau$. The $\phi E/\Delta\tau$ is negligibly small at the four corners in the figures, and therefore they may be taken as the reference points.

RIGOROUS SOLUTION TO THE EXTENDED KINETIC EQUATIONS FOR HOMOGENEOUS GAS MIXTURES

Giampiero Spiga
Department of Mathematics, University of Bari,
Via G.Fortunato, 70125 BARI, Italy.

Abstract

The paper deals with a recently introduced extended kinetic theory for gas mixtures. Such a generalization is aimed at accounting for effects of removal, generation and source, as well as for the presence of a background medium. These physical effects are modeled by suitable gain and loss terms into the set of Boltzmann equations governing the gas distribution functions. The outstanding work done for the search of exact or rigorous analytical solutions to the standard Boltzmann equation in the space homogeneous case and for Maxwell scattering models is reviewed. It is then shown how the previous results can be generalized to treat also gas mixtures in the extended version. In particular, in spite of the nonlinearities, a Fourier transform technique, with suitably defined generating functions, leads to a hierarchy of moment equations solvable in cascade, and a series reconstruction of the distribution functions, converging in an appropriate Hilbert space, follows then by resorting to a Bobylev transformation. Finally, the role played by the theory of dynamical systems in extended kinetic theory is discussed, and some examples of application are presented and commented on.

1. Introduction

According to a research line started at the Nuclear Engineering Laboratory of the University of Bologna, removal and creation inter-particle interactions (chemical or nuclear reactions, including self-generation), as well as the presence of a background medium (bath) and of external sources, can be taken into account in kinetic theory by a suitable modeling of their effects on the overall balance equations in the phase space. Under the well known assumptions of standard kinetic theory [14,54], (in particular, no encounters of order greater than two), the proper investigation tool is a set of "extended" nonlinear transport equations for a mixture of gases, with generalized gain and loss terms added to the usual scattering-in and scattering-out collision terms, which in turn are expressed according to the Boltzmann equation [46]. Mathematical and physical problems related to the above subject have been investigated in recent years by several authors in Italy as well as in the United States, in the Federal Republic of Germany and in Argentina [9,50,10,47,6,32,41,20,43,21,59,58]. The present notes are mainly aimed at devising and discussing analytical methods for determining the gas distribution functions from the previous set of integrodifferential Boltzmann-like equations. Therefore, practical issues (mainly the search for exact particular closed form solutions or the set-up of a rigorous constructive algorithm for representing the solution in a general situation) will be addressed to in the sequel, rather than more theoretical aspects, like well posedness of the abstract mathematical problem. More precisely, some of the outstanding work of the last decade on exact solutions to the nonlinear space-homogeneous and isotropic Boltzmann equation will be generalized to the extended case. The paper is organized as follows. The extended kinetic equations are first recalled and discussed on both mathematical and physical grounds in Section 2. Then, a brief review of exact or rigorous solutions to the Boltzmann equation in classical kinetic theory is given in Section 3. Section 4 deals with the generalization of a previously described moment method to the extended transport equations, and with a Laguerre series solution for the distribution functions of the considered homogeneous gas mixture. Finally, two simple examples of application are presented and shortly commented on in Section 5, mainly in order to show that a quite complicated dynamics, even a chaotic dynamics, may arise in extended kinetic theory.

2. Extended set of Boltzmann equations

The set of Boltzmann transport equations governing the distribution functions $f_i(\underline{x},\underline{v},t)$ of a gas of N species interacting between themselves via elastic scattering collisions reads as (in standard notation)

$$\frac{\partial f_i}{\partial t} + \underline{v}\cdot\overline{\nabla}_{\underline{x}} f_i + \frac{1}{m_i}\underline{F}_i\cdot\overline{\nabla}_{\underline{v}} f_i = \sum_{j=1}^{N} J^S_{ij}(f_i,f_j) + S_i \qquad i=1,2,\ldots,N \tag{1}$$

where m,F,S stands for mass, external force, and external source (or sink), respectively. The most general form for the scattering collision term J^S_{ij} is [56] ($\underline{x}$ and t dependence in the f_i is omitted in the remainder of this section for brevity, and integrations are understood to extend all over R_3)

$$J_{ij}^{S} = \iiint W_{ij}(\underline{v}',\underline{w}'\to\underline{v},\underline{w})[f_i(\underline{v}')f_j(\underline{w}')-f_i(\underline{v})f_j(\underline{w})]d\underline{v}'d\underline{w}'d\underline{w} \tag{2}$$

where W_{ij} denotes the probability density of a collision which carries particles i and j from velocities $\underline{v}',\underline{w}'$ into $\underline{v},\underline{w}$, and where the 9-fold integral actually collapses to a 5-fold one because of the four constraints imposed by conservation of momentum and energy. It has been shown [46] that two equivalent forms originate from Eq.(2), namely the standard collision integral of kinetic theory [14], that will be used later on, and the so-called scattering kernel formulation [7,51]. In the frame of the latter, one may introduce quantities of clear physical meaning, like the microscopic scattering collision frequency

$$g_{ij}^{S}(|\underline{v}-\underline{w}|) = \iint W_{ij}(\underline{v}',\underline{w}'\to\underline{v},\underline{w})d\underline{v}'d\underline{w}', \tag{3}$$

and the scattering kernel

$$\pi_{ij}(\underline{v}',\underline{w}'\to\underline{v}) = \frac{1}{g_{ij}^{S}(|\underline{v}'-\underline{w}'|)}\int W_{ij}(\underline{v}',\underline{w}'\to\underline{v},\underline{w})d\underline{w}. \tag{4}$$

π_{ij} represents the velocity probability distribution of a scattered particle i, originally at velocity $\underline{v}'$, after a collision with a particle j at velocity $\underline{w}'$, and its integral with respect to $\underline{v}$ is normalized to unity.

The presence of a background medium can be accounted for by adding a component (labeled below by the index o) interacting with all other species, with a distribution function f_o known a priori and independent of the process under examination. This leads to linear collision terms of the kind

$$J_{io}^{S} = -\hat{g}_i^{S}(|\underline{v}|)f_i(\underline{v}) + \int \hat{g}_i^{S}(|\underline{v}'|)\hat{\pi}_i(\underline{v}'\to\underline{v})f_i(\underline{v}')d\underline{v}' \tag{5}$$

where

$$\hat{g}_i^{S}(|\underline{v}|) = \int g_{io}^{S}(|\underline{v}-\underline{w}|)f_o(\underline{w})d\underline{w} \tag{6}$$

denotes the macroscopic scattering collision frequency for a particle i with a field particle, while

$$\hat{\pi}_i(\underline{v}'\to\underline{v}) = \frac{1}{\hat{g}_i^{S}(|\underline{v}'|)}\int g_{io}^{S}(|\underline{v}'-\underline{w}'|)\pi_{io}(\underline{v}',\underline{w}'\to\underline{v})f_o(\underline{w}')d\underline{w}' \tag{7}$$

is the corresponding scattering kernel; again its integral with respect to $\underline{v}$ is normalized to one.

Removal (absorption) of particles i by any kind of binary interactions with particles j can be easily modeled by adding a removal collision term to the rhs of Eq.(1), of the same form as the scattering-out contribution, namely

$$J^A_{ij}(f_i,f_j) = - f_i(\underline{v}) \int g^A_{ij}(|\underline{v}-\underline{w}|) f_j(\underline{w}) d\underline{w} \qquad \begin{matrix} i=1,\dots,N \\ j=0,1,\dots,N \end{matrix} \tag{8}$$

where g^A_{ij} is the collision frequency for all removal encounters. If the interaction gives rise eventually to any participating species, a gain term related to Eq.(8) will appear in some of the equations of the set (1), as described below, whereas, if no participating species are produced, the interaction yields only a loss term of the type (8). In the latter case the evolution of the generated particles is decoupled from the main problem, and may be determined independently "a posteriori". The symbol g^A_{ij} is comprehensive of all interaction mechanisms between species i and j making particles i disappear; the part of g^A_{ij} describing the passive interactions not leading to generation of participating species will be denoted by g^R_{ij}.

Creation collisions, in which particles i are generated by interaction between two other species (binary chemical or nuclear reaction) or by self-generating process (like fission for neutrons or ionization for electrons), can correspondingly be modeled and included into the picture. It is sufficient to write, in the rhs of Eq.(1), an additional creation collision term in the i-th equation as

$$\sum_{h=0}^{N} \sum_{j=h}^{N} \iint g^C_{hj,i}(|\underline{v}'-\underline{w}'|) \chi^i_{hj}(\underline{v}',\underline{w}'\to\underline{v}) f_h(\underline{v}') f_j(\underline{w}') d\underline{v}' d\underline{w}' \tag{9}$$

where $g^C_{hj,i}=g^C_{jh,i}$ is the microscopic frequency of collision between a particle h of velocity $\underline{v}'$ and a particle j of velocity $\underline{w}'$ producing particle i, while $\chi^i_{hj} d\underline{v}$ is the number of such particles that are given a final velocity in $d\underline{v}$ at $\underline{v}$. The mean number

$$\eta^i_{hj}(\underline{v}',\underline{w}') = \int \chi^i_{hj}(\underline{v}',\underline{w}'\to\underline{v}) d\underline{v} \tag{10}$$

of secondary particles i produced by collision between species h (at velocity $\underline{v}'$) and j (at velocity $\underline{w}'$) will also be useful later on: it may even be greater than unity. Notice that selfgeneration corresponds to either one of the options h=i or j=i, and that, for clear physical reasons, $g^C_{ii,i}$ is equal to zero; in addition, $g^C_{ij,j}$ must vanish as soon as $g^C_{ij,i}>0$. If all creation interactions produce only one kind of participating species, then one may write

$$g^A_{ij} = g^R_{ij} + \sum_{h=1}^{N} g^C_{ij,h}; \tag{11}$$

otherwise, weighting factors should be introduced in the sum on the rhs. Sometimes, it will be convenient to refer to the total microscopic collision frequency (relevant to all kind of interactions between species i and j) defined as

$$g_{ij} = g^S_{ij} + g^A_{ij}. \tag{12}$$

A few remarks are in order before concluding this section. Several physical phenomena can be described in the frame of the previous equations; for instance inelastic scattering and all processes with an intermediate compound state. Moreover, additional linear collision terms similar to (5) could be used to mimic disappearance and/or generation of particles i due to dissociation [8]; it would be sufficient to implement once more the rhs of Eq.(1) by inserting

$$- \lambda_i f_i(\underline{v}) + \sum_{j\neq i} \lambda_j \int \chi_j^i(\underline{v}'\rightarrow\underline{v}) f_j(\underline{v}')d\underline{v}' \qquad (13)$$

where λ_j is the decay constant for particles j, and $\chi_j^i d\underline{v}$ is the number of particles i emerging in $d\underline{v}$ at $\underline{v}$ as a result of dissociation of a particle j at velocity $\underline{v}'$. The inclusion of the effects (13) into the balance equations (1) would not add any significant difficulty to their solution (that is the main task undertaken here), and therefore will not be considered in the sequel. Finally, it must be noticed that the usual conservation laws of classical kinetic theory are lost in the present extended version. In fact, the total number of particles of any species is not constant, because of removals, creations and sources, and, even in absence of that, the average kinetic energy per particle is not conserved, because of energy exchange in elastic collisions with field particles. Because of the several, often competing, effects accounted for, the asymptotic trend is hardly predictable not only for the distribution functions, but even for the number densities

$$\rho_i(\underline{x},t) = \int f_i(\underline{x},\underline{v},t)d\underline{v} \ . \qquad (14)$$

In conclusion, a rather complicated dynamics has to be expected, and the investigation is made more difficult by the lack of some of the most useful classical tools, like conservation laws and H theorem.

3. Exact solutions to the classical Boltzmann equation

It seems out of question that the standard Boltzmann equation for a single gas, in its general form, will probably resist analytical solution for an indefinite (extremely long anyway) period of time; on the other hand there are physical problems for which it would be very useful to have at our disposal some exactly soluble models, in order to shed light on new aspects to be understood, and to obtain precious hints for the investigation of more general cases. For instance, approach to equilibrium is known to be highly nonuniform with respect to velocity, i.e. to the order of the velocity moments of the distribution function, and the need of more insight into this nonuniform relaxation process (controlled by the nonlinear term) motivated the pioneering work of Bobylev and Krook-Wu, who independently discovered the first analytical solution to the nonlinear Boltzmann equation [5,34]. At the same time, such exact solutions are also important as a benchmark for numerical schemes pretending to deal with more general situations. Models used in this context should be simple enough to allow analytical manipulations, but meaningful enough to represent physical conditions. Most of the literature on the subject is confined to problems which are homogeneous in space and isotropic in velocity. To the author's knowledge, very little has been done for the space inhomogeneous case [39,38,16], whereas the generalization to the angular dependent case, though technically heavy, is certainly possible [57,31,3]. For the sake of simplicity however, only an infinite homogeneous space domain with isotropic physical conditions will be dealt with here. As regards the interparticle collision law, there is little doubt that Maxwell models, related in different possible ways to an inverse 4-th power intermolecular potential, present great advantages, even though other models with analogous properties have been proposed [22]. In fact, collision frequencies and scattering kernels take reasonably simple expressions, the linearized collision operator, even though not compact, has a pure point spectrum, and its eigenfunctions are explicitly known (generalized Laguerre polynomials in v^2 times a Maxwellian). In particular, the model of Maxwell molecules with angular cutoff in three dimensions, to be employed here, guarantees [46] that scattering-in and scattering-out contributions are well defined separately, and that the scattering collision frequency g^S is a constant. Moreover, the differential scattering cross section, which contains all details of the encounter between two particles at velocities $\underline{v}'$ and $\underline{w}'$, reads as

$$\frac{1}{|\underline{v}'-\underline{w}'|} g^S \theta(\hat{\Omega}\cdot\hat{\Omega}') \tag{1}$$

where $\hat{\Omega}$ and $\hat{\Omega}'$ are the unit vectors in the direction of $\underline{v}-\underline{w}$ and $\underline{v}'-\underline{w}'$, $\underline{v}$ and $\underline{w}$ are the velocities after collision, and the angular distribution θ is normalized as

$$\int_{4\pi} \theta(\hat{\Omega}\cdot\hat{\Omega}')d\hat{\Omega}' = 1 \ . \tag{2}$$

Under the previous assumptions, the approach by Krook and Wu is based on the moment method, which is of course not new in kinetic theory [29,33]. Introduce moments

$$M_n(t) = \frac{1}{(2n+1)!!} \int v^{2n} f(v,t) d\underline{v} \tag{3}$$

and generating function

$$G(z,t) = \sum_{n=0}^{\infty} M_n(t) z^n \ , \tag{4}$$

supposed existing, with $M_0=\rho$, and $M_1=KT/m$ related to the energy (K is the Boltzmann constant). Taking moments of the Boltzmann equation yields, after some algebra, the set of moment equations, in dimensionless variables,

$$\frac{dM_n}{d\tau} + M_n(\tau) = \frac{1}{n+1} \sum_{p=0}^{n} M_p(\tau) M_{n-p}(\tau) \qquad n=2,3,\ldots \tag{5}$$

with M_0=const. and M_1=const. (conservation laws), and the nonlinear partial differential equation for G

$$\left[\frac{\partial^2}{\partial z \partial \tau} + \frac{\partial}{\partial z} \right] [zG(z,\tau)] = G^2(z,\tau) \tag{6}$$

for which it was possible to find a similarity solution. Then, by a miracolous sequence of analytical steps, the impressive task was accomplished of obtaining the distribution function f(v,t) starting from the generating function G; the procedure went through the evaluation of all M_n as derivatives of G, the sum of the power series representing the Fourier cosine transform of f (whose n-th coefficient is proportional to M_n), and the final Fourier inversion. The result is the famous BKW mode

$$f(v,t) = \rho \left[\frac{m(1+\sigma)}{2\pi KT} \right]^{3/2} \left\{ 1 + \sigma \left[\frac{m(1+\sigma)}{2KT} v^2 - \frac{3}{2} \right] \right\} \exp\left[- \frac{m(1+\sigma)}{2KT} v^2 \right] \tag{7}$$

where T=constant, $\rho=\rho^*$=constant, and $\sigma=\sigma^*\Phi(t)/\{1+\sigma^*[1-\Phi(t)]\}$, with

$$\Phi(t) = \exp(-\delta g^S \rho^* t) \qquad \delta = \frac{\pi}{2} \int_{-1}^{1} (1-\mu^2)\theta(\mu) d\mu \; ; \tag{8}$$

starred quantities refer to the initial conditions at t=0. Positivity requirements imply the constraint $0 \le \sigma^* \le 2/3$. Since $\sigma(t) \to 0$ for $t \to \infty$, Eq.(7) describes a relaxation to an absolute Maxwellian, starting from an initial state, which is a BKW mode at $\sigma=\sigma^*$.

The BKW mode has been generalized in different ways. If particles are allowed to remove each other, Eq.(7) is still in order, but with [48]

$$\rho(t) = \rho^*(1+g^R\rho^*t)^{-1} \qquad \Phi(t) = (1+g^R\rho^*t)^{-\delta g^S/g^R}, \tag{9}$$

namely the number density vanishes asymptotically, while the shape function $f(v,t)/\rho(t)$ tends to the same Maxwellian as before, but in a nonexponential fashion. A purely removing background can be accounted for as well [11,37]; it is sufficient to modify the functions ρ and Φ as

$$\rho(t) = \hat{g}^R\rho^*[(\hat{g}^R+g^R\rho^*)\exp(\hat{g}^R t) - g^R\rho^*]^{-1} \tag{10a}$$

$$\Phi(t) = \{1+g^R\rho^*[1-\exp(-\hat{g}^R t)]/\hat{g}^R\}^{-\delta g^S/g^R}, \tag{10b}$$

with only the new important feature that the asymptotic shape function is not a Maxwellian, but another BKW mode.

A different kind of generalization concerns the simple addition of an external source [53,40,44]. By extensively using Lie group theoretical methods, families of exact particular solutions corresponding to special choices of the source have been determined. Positivity preserving arguments play a fundamental role in selecting solutions of physical meaning, especially if the source is negative. In some more general cases, the whole procedure still works, except the last step (Fourier inversion) that needs to be performed numerically [45].

All previous solutions are only particular integrals to the transport equation, and can not satisfy a given arbitrary initial condition. Moreover, they can be determined only after many different steps, each one with its own difficulties, which must be overcome analytically, so that the long chain is very likely to break down somewhere. There is thus a need for rigorous solutions, by which arbitrary initial conditions can be met, and the distribution can be reconstructed by a sequence of approximate solutions converging to the exact one in a suitable norm. In this respect, bearing in mind that Laguerre polynomials in the energy variable (ranging from 0 to ∞) are essentially the eigenfunctions of the linearized collision operator, the most direct and simple idea of a reconstruction in terms of a Laguerre series arises spontaneously. For this purpose one should evaluate the Laguerre moments ξ_n of the distribution function, linearly related to the power moments M_n, that follow in turn from the solution of the set of equations (5). Such moment equations have at least two remarkable properties: first, in spite of the nonlinearity of the problem, each equation is linear in its unknown, and secondly, they can be solved in cascade, for the n-th equation contains moments only up to n. However, since evaluating Laguerre moments from the computed ordinary moments would be subject to numerical instability, the Laguerre moments must be determined directly and independently.

For the whole procedure to work, one has to assume that f belongs to the real Hilbert space H with norm (in dimensionless variables) [17]

$$\int \exp(v^2/2)|f(v)|^2 d\underline{v} < \infty \;, \tag{11}$$

that ensures in particular existence of all moments ξ_n and M_n. The eigenfunctions of the linearized collision operator in H, to be used as orthogonal expansion basis, read then as $\exp(-v^2/2)L_n^{(1/2)}(v^2/2)$. It has been observed that the generating function may be considered as an integral transform of the distribution function [18]

$$G(z,\tau) = \int {}_1F_1(1,\frac{3}{2};\frac{zv^2}{2})\; f(v,\tau)d\underline{v} \tag{12}$$

where ${}_1F_1$ denotes confluent hypergeometric function. From the expansion

$$f(v,\tau) = \exp(-v^2/2) \sum_{n=0}^{\infty} \xi_n(\tau)L_n^{(1/2)}(v^2/2) \tag{13a}$$

it can be shown that

$$G(z,\tau) = \sum_{n=0}^{\infty} \xi_n(\tau) \frac{z^n}{(1-z)^{n+1}} \;, \tag{13b}$$

with isolated essential singularity at z=1, which implies a finite radius of convergence for Eq.(4). Inserting the expansion (13b) into the nonlinear PDE (6), it has been proved [15] that the proper ordering in terms of rational functions of z yields, for the unknowns $\xi_n(\tau)$, exactly the same set (5) of moment equations, with $\xi_0=\rho=$constant, and $\xi_1=$constant$=0$. Laguerre moments are thus again determined by solving sequentially linear ODE's, and differ from the moments M_n only because of the different initial conditions to be employed. Once the ξ_n are known, the distribution function is provided, for any initial data, by the series (13a), converging in the norm of H. The same mathematical tool has been used in order to single out exact particular solutions corresponding to a finite number of Laguerre moments different from zero [2]. Of course, only solutions which remain nonnegative starting nonnegative may be considered in the latter class.

However, the norm (11) has no physical meaning, and even the conjecture that it might be related to the finitess of entropy turned out to be not convincing [55]. On the other hand, it is very convenient for the application of the theory of self-adjoint operators in Hilbert spaces. Even though conditions have been investigated for f to belong to H if the initial condition belongs to H, and for f to tend to a Maxwellian when $t\to\infty$, the solution f does not need to belong to H in general [15]. A larger Hilbert space, and solutions of the nonlinear Boltzmann equation which decrease like inverse powers of the energy have been introduced [18,60]. Only a finite number of moments exists for them. They are associated to the powerlike decreasing eigenfunctions of the linearized collision operator belonging to the continuous spectrum, and will not be considered here.

4. Laguerre series solution to the extended kinetic equations

In this section, some of the results recalled above about rigorous solutions in kinetic theory will be generalized to the extended problem described in Sec.2, keeping the same assumptions of homogeneity in space, isotropy in velocity, and Maxwell-like scattering model. In agreement with that, all collision frequencies of any kind will be taken to be constant, and all χ's to be independent of $\underline{v}'$ and $\underline{w}'$, so that the kinetic equations to be dealt with read as

$$\frac{\partial f_i}{\partial t} + \sum_{j=0}^{N} g_{ij}\rho_j(t)f_i(v,t) = S_i(v,t) + \sum_{h=0}^{N}\sum_{j=h}^{N} g^C_{hj,i}\rho_h(t)\rho_j(t)\chi^i_{hj}(v) + \sum_{j=0}^{N} g^S_{ij}\int d\underline{w}\int_{4\pi}\theta_{ij}(\hat{\Omega}\cdot\hat{\Omega}')\, f_i(v',t)f_j(w',t)d\hat{\Omega}' \tag{1}$$

$i=1,2,\ldots,N$, where

$$\underline{v}' = (r_{ij}\underline{v} + \underline{w} + V\hat{\Omega}')/(r_{ij} + 1) \qquad \underline{w}' = (r_{ij}\underline{v} + \underline{w} - r_{ij}V\hat{\Omega}')/(r_{ij} + 1)$$
$$V = |\underline{v} - \underline{w}| = |\underline{v}' - \underline{w}'| \qquad \hat{\Omega} = (\underline{v} - \underline{w})/V \tag{2}$$

and r_{ij} denotes the mass ratio m_i/m_j. Loss terms are all combined in the second term in the lhs of Eq.(1), while the gain terms are divided into external, creation and scattering-in contributions in the rhs. Since the densities are integrals of the unknowns, nonlinearities of separable type appear, in addition to the nonlinear integro-functional nature of the last addend. Solutions f_i belonging to the real weighted Labesgue spaces $L_2^{\beta_i}$ defined by the norm

$$\int \exp(\beta_i v^2)\ |f_i|^2\ d\underline{v} < \infty \tag{3}$$

will be sought; notice that the number $\beta_i>0$ is not uniquely determined. Similar hypotheses will be made for S_i, χ^i_{hj} and for the initial data $f_i(v,0)$. The assumption guarantees existence of all moments (3.3), as well as of the Fourier transforms

$$\tilde{f}_i(\underline{k},t) = \int f_i(v,t)\exp(-\underset{\sim}{i}\underline{k}\cdot\underline{v})\,d\underline{v} = F[f_i(v,t)] = \tilde{f}_i(k,t)\ , \tag{4}$$

where $\underset{\sim}{i}$ is the imaginary unit, and

$$f_i(v,t) = \frac{1}{(2\pi)^3}\int \tilde{f}_i(k,t)\exp(\underset{\sim}{i}\underline{k}\cdot\underline{v})\,d\underline{k} = F^{-1}[\tilde{f}_i(k,t)]\ . \tag{5}$$

Even though Fourier transform is essentially a linear tool, a Fourier transform approach turned out to be very effective for the classical nonlinear Boltzmann equation [4]. In the present extended context, a rearrangement of integrations in the set of transformed equations, made possible by the conservation laws in a scattering collision, and by the Liouville theorem $d\underline{v}d\underline{w} = d\underline{v}'d\underline{w}'$ [14], leads to the nonlinear equations

$$\frac{\partial\tilde{f}_i}{\partial t} + \sum_{j=0}^{N} g_{ij}\rho_j(t)\tilde{f}_i(k,t) = \tilde{S}_i(k,t) + \sum_{h=0}^{N}\sum_{j=h}^{N} g^C_{hj,i}\rho_h(t)\rho_j(t)\tilde{\chi}^i_{hj}(k) +$$
$$+ \sum_{j=0}^{N} g^S_{ij}\iint f_i(v',t)f_j(w',t)\exp\left[-\underset{\sim}{i}\underline{k}\cdot\frac{r_{ij}\underline{v}'+\underline{w}'}{r_{ij}+1}\right]d\underline{v}'d\underline{w}'\int_{4\pi}\theta_{ij}(\hat{\Omega}\cdot\hat{\Omega}')\exp\left[-\underset{\sim}{i}\underline{k}\cdot\frac{V\hat{\Omega}}{r_{ij}+1}\right]d\hat{\Omega}. \tag{6}$$

Now, setting $\underline{k}=k\hat{k}$, with $|\hat{k}|=1$, one can realize that the last integral with respect to $\hat{\Omega}$ actually depends on $\underline{k}$, V and $\hat{\Omega}'$ only through the combinations kV and $\hat{k}\cdot\hat{\Omega}'$, so that the roles of $\hat{k}$ and $\hat{\Omega}'$ may be interchanged. Therefore, since $V\hat{\Omega}'=\underline{v}'-\underline{w}'$, a permissible inversion of integration order allows to cast also the last integral in terms of the transforms $\tilde{f}_i$ and $\tilde{f}_j$; more precisely, since the transforms themselves are isotropic, one ends up with the selfcontained set of differential functional equations

$$\frac{\partial \tilde{f}_i}{\partial t} + \sum_{j=0}^{N} g_{ij}\rho_j(t)\tilde{f}_i(k,t) = \tilde{S}_i(k,t) + \sum_{h=0}^{N}\sum_{j=h}^{N} g^C_{hj,i}\rho_h(t)\rho_j(t)\tilde{\chi}^i_{hj}(k) +$$

$$+ 2\pi \sum_{j=0}^{N} g^S_{ij} \int_{-1}^{1} \theta_{ij}(\mu)\tilde{f}_i\left[k\,\frac{(r_{ij}^2+1+2r_{ij}\mu)^{1/2}}{r_{ij}+1},t\right]\tilde{f}_j\left[k\,\frac{(2-2\mu)^{1/2}}{r_{ij}+1},t\right]d\mu \tag{7}$$

with a one dimensional integral in the scattering collision term, instead of the five fold one of Eq.(1). Now, moment equations are most easily derived by generalizing the elegant procedure of Ref.22. If the Fourier transforms are rewritten as

$$\tilde{f}_i(k,t) = 4\pi \int_0^\infty f_i(v,t)\,\frac{\sin kv}{kv}\,v^2 dv = \sum_{n=0}^{\infty} \frac{(-1)^n}{n!} M^i_n(t)\,(k^2/2)^n, \tag{8}$$

with $M^i_0(t) = \rho_i(t)$ and

$$M^i_n(t) = \frac{1}{(2n+1)!!}\int v^{2n} f_i(v,t)\,d\underline{v}\,, \tag{9}$$

it is readily seen that $\tilde{f}_i$ may be regarded as a sort of generating function of the moments with parameter $k^2/2$, defined by a series with an infinite radius of convergence (in other words, $\tilde{f}_i$ is an entire function of k). One can write analogously

$$\tilde{S}_i(k,t) = \sum_{n=0}^{\infty} \frac{(-1)^n}{n!} S^i_n(t)\,(k^2/2)^n \qquad \tilde{\chi}^i_{hj}(k) = \sum_{n=0}^{\infty} \frac{(-1)^n}{n!} Q^{hj,i}_n (k^2/2)^n\,, \tag{10}$$

and then, upon inserting Eqs.(8) and (10) into Eq.(7), it is possible to factor out the integrals with respect to μ, and to order in powers of k, obtaining after some algebra, the moment equations

$$\frac{dM^i_n}{dt} + M^i_n(t)\sum_{j=0}^{N} g_{ij}M^j_0(t) = S^i_n(t) + \sum_{h=0}^{N}\sum_{j=h}^{N} g^C_{hj,i}Q^{hj,i}_n M^h_0(t)M^j_0(t) +$$

$$+ \sum_{j=0}^{N} g^S_{ij} \sum_{p=0}^{n} \mu^{ij}_{np}(r_{ij})M^i_p(t)M^j_{n-p}(t) \tag{11}$$

$i=1,\dots,N$; $n=0,1,\dots$, with coefficients

$$\mu^{ij}_{np}(r) = \binom{n}{p}\frac{2^n r^p}{(r+1)^{2n}}\,2\pi \int_{-1}^{1}\theta_{ij}(\mu)\left[\frac{r^2+1}{2r}+\mu\right]^p (1-\mu)^{n-p}d\mu \tag{12}$$

depending on the mass ratios r_{ij}, which are a measure of the energy exchange in a scattering collision. In particular $\mu^{ij}_{oo}(r)=1$ for any i,j,r, and the standard coefficients [22] are recovered for i=j and r=1. Analytical expressions become easily available once θ_{ij} is given; for instance, in the simplest case of the Krook-Wu model, namely $\theta_{ij}(\mu)=1/4\pi$, one gets

$$\mu_{np}(r) = \frac{1}{n+1}\left[\frac{2}{1+r}\right]^{2n}\sum_{h=0}^{p}\binom{n+1}{h}\left[\frac{1-r}{2}\right]^{2h} r^{p-h} \tag{13}$$

with, in particular

$$\mu_{np}(1) = \frac{1}{n+1} \qquad \mu_{nn}(r) = \frac{1}{n+1}\,\frac{1-\alpha^{n+1}}{1-\alpha}, \tag{14}$$

where $\alpha=[(1-r)/(1+r)]^2$. In any event, it has been proved that moment equations can be generalized to treat a mixture of different species in extended kinetic theory. New features of Eq.(11) are that M_0^i and M_1^i are not constant (lack of conservation laws), and thus equations at levels n=0 and n=1 must also be considered. For n=0 one recovers the continuity equations, a set of selfcontained 1st order semilinear ODE with quadratic nonlinearities for the number densities. These equations carry outstanding physical significance by themselves, and have been considered already for some applications of practical interest [12,13,49]. For n>0, the equations from the set (11) are linear, and retain the nice feature of solvability in cascade.

The first few moments are often sufficient for practical purposes, and Eq.(11) provides a very convenient tool for their evaluation. But power moments are useless for a reconstruction of the distribution functions. The proper orthogonal expansion basis in the Hilbert space $L_2^{\beta_i}$ is provided by the denumerable set of eigenfunctions in $L_2^{\beta_i}$ of the linearized collision operator J_{ii}^S, which belong to the discrete scaled eigenvalues -n/(n+1). If M stands for the normalized Maxwellian

$$M(v,\beta_i) = \left[\frac{\beta_i}{\pi}\right]^{3/2} \exp(-\beta_i v^2) \qquad \beta_i = \frac{m_i}{2KT_i}, \tag{15}$$

the general n-th eigenfunction reads as

$$M(v,\beta_i)\, L_n^{(1/2)}(\beta_i v^2) \qquad n = 0,1,\ldots \tag{16}$$

and Laguerre moments and Laguerre series are defined by

$$\xi_n^i(t) = \frac{2^n n!}{(2n+1)!!}\int f_i(v,t) L_n^{(1/2)}(\beta_i v^2)\, d\underline{v} \tag{17}$$

and

$$f_i(v,t) = M(v,\beta_i) \sum_{n=0}^{\infty} \xi_n^i(t) L_n^{(1/2)}(\beta_i v^2). \tag{18}$$

In addition, $\xi_0^i(t)=\rho_i(t)$, and

$$\xi_n^i(t) = \sum_{p=0}^{n}\binom{n}{p}(-1)^p(2\beta_i)^p M_p^i(t) \qquad M_n^i(t) = \frac{1}{(2\beta_i)^n}\sum_{p=0}^{n}\binom{n}{p}(-1)^p\xi_p^i(t). \tag{19}$$

Notice that the Maxwellian $M(v,\beta_i)$ would give rise to moments $\xi_n^i=\delta_{no}$ (Kronecker symbol). Laguerre moments of the functions $S_i(v,t)$ and $\chi_{hj}^i(v)$ are analogously defined as

$$\bar{S}_n^i(t) = \frac{2^n n!}{(2n+1)!!} \int S_i(v,t)\, L_n^{(1/2)}(\beta_i v^2) d\underline{v}$$

$$\bar{Q}_n^{hj,i} = \frac{2^n n!}{(2n+1)!!} \int \chi_{hj}^i(v)\, L_n^{(1/2)}(\beta_i v^2) d\underline{v} \; . \tag{20}$$

The idea that enables one to derive now moment equations for ξ_n^i is a generalization of the concept of Bobylev transformation, which is expressed by the ansatz [22]

$$\tilde{f}_i(k,t) = \tilde{M}(k,\beta_i)\, \Phi_i(k,t) = \exp(-k^2/4\beta_i)\, \Phi_i(k,t) \; . \tag{21}$$

The classical Boltzmann equation is in fact invariant under Bobylev transformation, and the Maxwellian represents equilibrium distribution. Now equilibrium (under scattering only) is expected when the distribution functions are Maxwellians at the same temperature, namely with parameters β_i related by

$$\beta_i/\beta_j = r_{ij} \,, \qquad i=1,\ldots,N; \quad j=0,1,\ldots N. \tag{22}$$

If the background temperature is defined, the parameter β_0 is known, and the relation (22) determines uniquely all β_i with $i\geq 1$. Otherwise, one degree of freedhom is left in the choice of such parameters. The role played by Eq.(22) is immediately recognized for, in order to get functional equations for the Φ_i, the ansatz (21) has to be inserted into the Fourier transformed kinetic equations (7). It is a matter of standard algebra to prove that, if betas are just chosen according to Eq.(22), then the exponential $\exp(-k^2/4\beta_i)$ cancels out from the i-th equation, and the Φ_i obey the same set of equations as the $\tilde{f}_i$, only with $\tilde{S}_i$, $\tilde{\chi}_{hj}^i$ and initial conditions replaced by the corresponding quantities under Bobylev transformation. Moreover, since $\tilde{f}_i$ is an entire function of k, with derivatives $\tilde{f}^{(2n)}(0)=(-1)^n(2n-1)!!M_n^i(t)$ and $\tilde{f}^{(2n+1)}(0)=0$, Φ_i is also an entire function of k, i.e.

$$\Phi_i(k,t) = \sum_{n=0}^{\infty} \frac{2^n}{(2n)!} \Phi_i^{(2n)}(0)\, (k^2/2)^n, \tag{23}$$

which, combined with Eq.(21), yields

$$\tilde{f}_i(k,t) = \sum_{n=0}^{\infty} \frac{1}{n!(2n-1)!!} \Phi_i^{(2n)}(0)\, (k^2/2)^n \exp(-k^2/4\beta_i) \; . \tag{24}$$

Eq.(24) may be inverted term by term, and, since

$$F^{-1}[(k^2/2)^n \exp(-k^2/4\beta_i)] = n!(2\beta_i)^n M(v,\beta_i) L_n^{(1/2)}(\beta_i v^2), \tag{25}$$

comparison to Eq.(18) determines $\Phi_i^{(2n)}(0)$ as $\frac{(2n-1)!!}{(2\beta_i)^n} \xi_n^i(t)$, and therefore the functions Φ_i are completely identified as

$$\Phi_i(k,t) = \sum_{n=0}^{\infty} \frac{1}{n!(2\beta_i)^n} \xi_n^i(t)\, (k^2/2)^n, \tag{26}$$

namely as a kind of generating function of the Laguerre moments, with parameter $k^2/2$; the series has again an infinite radius of convergence. The final step in order to get moment equations for ξ_n^i consists in plugging the expansion (26) into the nonlinear functional equations for Φ_i, and ordering in powers of k. Omitting all details, the following set of moment equations is derived

$$\frac{d\xi_n^i}{dt} + \xi_n^i(t) \sum_{j=0}^{N} g_{ij}\xi_0^j(t) = \bar{S}_n^i(t) + \sum_{h=0}^{N} \sum_{j=h}^{N} g_{hj,i}^{C} \bar{Q}_n^{hj,i} \xi_0^h(t)\xi_0^j(t) +$$

$$+ \sum_{j=0}^{N} g_{ij}^{S} \sum_{p=0}^{n} r_{ij}^{n-p} \mu_{np}^{ij}(r_{ij})\xi_p^i(t)\xi_{n-p}^j(t) \qquad (27)$$

i=1,...,N; n=0,1,..., which is, mutatis mutandis, the same as Eq.(11), apart from the factor r_{ij}^{n-p} in the scattering term. The most important features of the set (11) are conserved. The only nonlinear step is the first one, at n=0, and corresponds once more to the continuity equations for the densities. All subsequent moments follow from linear first order ODE's, sequentially solvable in recursive form. Notice that, if Laguerre moments are evaluated exactly, convergence of each of the series (18), i=1,...,N, to the exact i-th distribution function is guaranteed a priori in the norm of $L_2^{\beta_i}$. Needless to say, the competing effects of scattering, host medium, removals, creations and sources on densities and shape functions may produce very easily quite complicated time behaviors. Numerical work is in progress [25,26], but the situation is sufficiently demonstrated by considering only the zero-th order equations, which deserve considerable attention by their own, because of their physical meaning, and also because all higher moments will be linearly determined in terms of the densities. Such equations read as

$$\dot{\rho}_i = - \sum_{j=0}^{N} g_{ij}^{A}\rho_i(t)\rho_j(t) + \sum_{h=0}^{N} \sum_{j=h}^{N} \eta_{hj}^{i} g_{hj,i}^{C}\rho_h(t)\rho_j(t) + \bar{S}_o^i(t) \qquad (28)$$

i=1,...,N, and are an autonomous set of ODE for the densities when the S_o^i are time independent. If this is the case, they give rise to a dynamical system, similar to those arising in several different fields (chemistry [42], biology [27], hydrodynamics [24], economics [28], and others), which often entail very complicated and fascinating asymptotic dynamics [19].

5. Two simple examples

In this section, two simple examples of application of Eq.(4.28), which have been object of some investigation in the literature, will be presented and briefly discussed, in order to show how the nonlinear interactions, that would be absent in the classical case, may lead to asymptotic trends that are completely unusual in transport theory, and even to deterministic chaos. Of course, these conclusions are not surprising, because of the well known nature of Eq.(4.28) itself, but the scientific community should be aware of them, since effects of removals, creations, sources and background media are usually considered only in linear transport, where the previous interesting effects may not be observed, since they are just due to the nonlinearities.

The first example refers to the sourceless case when creation reduces to selfgeneration only, namely $g^C_{hj,i}=0$ unless h=i or j=i [6,32]. Setting

$$C_{ij} = g^R_{ij} + g^C_{ij,j} - (\eta^i_{ij} - 1)g^C_{ij,i} \qquad \nu_i = C_{io}\rho_o \tag{1}$$

one ends up with the Lotka-Volterra system

$$\dot{\rho}_i = - \rho_i(t) \left[\sum_{j=1}^{N} C_{ij}\rho_j(t) + \nu_i \right] \qquad i=1,2,\dots,N. \tag{2}$$

All dependent variables are nonnegative, and the admissible phase space is the closure $\bar{R}_N^+$ of the positive orthant of R_N, bounded by the invariant planes $\rho_i=0$. Time evolution is controlled by the signs of C_{ij}, that may be negative only for $\eta^i_{ij}>1$, corresponding to prevailing of regeneration over removal. The system (2) in the present physical context is subject to the constraints that follow from the assumptions on the collision frequencies, namely i) $C_{ii}\geq 0$, and ii) whenever $C_{ij}<0$, then $C_{ji}>0$. In some (unphysical) cases where the constraints above are relaxed, divergence of some of the densities in finite time, i.e. lack of a global solution, can be proved [12]. Several types of phase trajectories have been observed: attraction by a stable stationary solution (fixed point), closed orbit, corresponding to periodic solution, depending on the initial point (conservative trend), and escape to infinity. Sensitive dependence on initial data also occurs [13,32]: already for N=2, there are situations with two stable one-species equilibrium points, and with their basins of attraction separated by the stable manifold of the unstable two species equilibrium point. Therefore, initial points very near by can end-up very far apart, with exhaustion of one species or the other. Typically, equilibrium points with more than one species enter the phase space by transcritical bifurcation with a lower order equilibrium point when parameters C_{ij} are varied. If the trace of C_{ij} is different from zero, the divergence of the vector field in the rhs of Eq.(2) with respect to the new variables $y_i=\ln(\rho_i/\rho^o_i)$ is negative. The same condition ensures existence of trapping bounded regions, so that dynamical structures which are typical of dissipative dynamical systems [30] and that have been found already in Lotka-Volterra systems, may take place [1]. There are for instance [36] values of the parameters for which a route to a chaotic dynamics originates from a supercritical Hopf bifurcation of the N-species equilibrium point, with transfer of stability to a limit cycle (periodic attractor). Then, a sequence of period-doubling bifurcations occurs when one parameter is varied, giving rise to cycles of period two, four, and so on, until a chaotic attractor develops [23].

In the second example, constant external sources in a three-component mixture are allowed, and two kinds of creation interactions are considered, one of which is not of selfgeneration type. More precisely, species 1 can be removed by all other species, including background, and be generated by fission-like collisions with species 2, and species 2 can be created by binary nuclear reactions between species 1 and 3 with $\eta^2_{13}=1$. Continuity equations read as

$$
\begin{aligned}
\dot\rho_1 &= (\eta^1_{12} g^C_{12,1} - g^A_{12})\rho_1\rho_2 - g^A_{13}\rho_1\rho_3 - \hat g^A_1\rho_1 + S^1_o \\
\dot\rho_2 &= - g^A_{12}\rho_1\rho_2 + g^C_{13,2}\rho_1\rho_3 + S^2_o \qquad (3)\\
\dot\rho_3 &= - g^A_{13}\rho_1\rho_3 + S^3_o
\end{aligned}
$$

and the admissible phase space is again $\overline{R}^+_3$, which is positively invariant if all S^i_o are nonnegative. A system like (3) is suggested by the analysis of the breeding process, as it occurs in a fast nuclear reactor, under appropriate assumptions and with suitable physical meaning for the parameters [52]. Species 1, 2 and 3 should then be identified with neutrons, fissile nuclei, and fertile nuclei, respectively; interactions between neutrons and fissile material (fuel) produce energy and additional neutrons to keep the reactor working, while new fuel is produced by nuclear reactions between neutrons and fertile material, the latter being supplied from the exterior ($S^3_o>0$), in order to replace removed nuclei. The case $S^1_o=0$ has been investigated already, but a positive neutron source is required, if one wishes to take free neutrons traveling in the atmosphere into account. Anyway, upon adimensionalizing dependent and independent variables, the system (3) may be rewritten as

$$
\begin{aligned}
\dot x_1 &= Ax_1x_2 - x_1x_3 - x_1 + E \\
\dot x_2 &= - Dx_1x_2 + x_1x_3 + F \qquad (4)\\
\dot x_3 &= - x_1x_3 + G
\end{aligned}
$$

where, for physical reasons, A,D,G are positive, and E is nonnegative, whereas F might be any real number; F<0 would simulate extraction of extra fuel from the reactor, but, in this case, Eq.(4) remains valid only as long as $x_2\geq 0$. The dynamics is quite dull for E=0, with most trajectories escaping to infinity tangentially to the invariant plane $x_1=0$, and with only one stationary solution, which exists and is stable in a certain domain of parameter space. However, such a fixed point is never a global attractor. It might undergo Hopf bifurcation and become unstable when parameters are varied, but the bifurcation turns out to be subcritical, and no stable periodic solution originates in this way [52]. More interesting effects are observed in the case E>0, that is being studied now, and will be object of a future work [35]. Again there is at most one fixed point, namely

$$\bar x_1 = \frac{B}{D} \qquad \bar x_2 = \frac{F+G}{B} \qquad \bar x_3 = \frac{DG}{B}, \qquad B = DE + AF + (A-D)G \qquad (5)$$

for

$$G > -F \qquad\qquad E > -\frac{A}{D}F + \frac{D-A}{D}G . \qquad (6)$$

No stationary solution is allowed when Eq.(6) is violated. The fixed point (5) is an asymptotically stable nonglobal attractor when E is large enough, and might loose stability by Hopf bifurcation when E is decreased. Now the observed bifurcation is actually supercritical, and gives rise to an attracting periodic solution, with oscillations changing from a regular sinusoidal-like shape, when E is still close to the bifurcation value and the limit cycle is almost an ellipsis surrounding the unstable fixed point, to a sequence of violent delta-like pulses with a very long period, when E has been decreased further, and the cycle has become large and irregular, approaching the x_1 axis on one side, and the x_1=0 plane on the other. However, this and other aspects of the dynamical system generated by (4) are still under investigation.

6. Acknowledgements

Work supported by the National Group for Mathematical Physics of the Italian Research Council (C.N.R.) in the frame of the activities PSMA-MMAI.

7. References

1. A.Arneodo,P.Coullet,J.Peyraud,C.Tresser, *Strange attractors in Volterra equations for species in competition*, J.Math.Biol.14, 153-157(1982).
2. M.Barnsley,H.Cornille, *On a class of solutions of the Krook-Tjon-Wu model of the Boltzmann equation*, J.Math.Phys.21, 1176-1193(1980).
3. R.O.Barrachina,C.R.Garibotti, *Nonisotropic solutions of the Boltzmann equation*, J.Stat.Phys.45, 541-560(1986).
4. A.V.Bobylev, *Fourier transform method in the theory of the Boltzmann equation for Maxwell molecules*, Sov.Phys.Dokl.20, 820-822(1976).
5. A.V.Bobylev, *Exact solutions of the Boltzmann equation*, Sov.Phys.Dokl.20, 822-824(1976).
6. V.C.Boffi,V.Franceschini,G.Spiga, *Dynamics of a gas mixture in an extended kinetic theory*, Phys.Fluids 28, 3232-3236(1985).
7. V.C.Boffi,V.G.Molinari, *Nonlinear transport problems by factorization of the scattering probability*, Nuovo Cimento 65B, 29-44(1981).
8. V.C.Boffi,A.Rossani, *On the Boltzmann system for a mixture of reacting gases*, to be published.
9. V.C.Boffi,G.Spiga, *Global solution to a nonlinear integral evolution problem in particle transport theory*, J.Math.Phys.23, 2299-2303(1982).
10. V.C.Boffi,G.Spiga,J.R.Thomas, *Solution of a nonlinear integral equation arising in particle transport theory*, J.Comp.Phys.59, 96-107(1985).

11. V.C.Boffi,G.Spiga, *Exact time dependent solutions to the nonlinear Boltzmann equations*, in Rarefied Gas Dynamics 15, V.Boffi and C.Cercignani Editors, Vol. 1, 55-63, Teubner, Stuttgart, 1986.

12. V.C.Boffi,G.Spiga, *Extended kinetic theory for gas mixtures in the presence of removal and regeneration effects*, Z.A.M.P.(J.Appl.Math.Phys.)37, 27-42(1986).

13. V.C.Boffi,G.Spiga, *Calculation of the number densities in an extended kinetic theory of gas mixtures*, Trans.Th.Stat.Phys.16, 175-188(1987).

14. C.Cercignani, Theory and Application of the Boltzmann Equation, Scottish Academic Press, Edinburgh, 1975.

15. H.Cornille, *On the Krook-Wu model of the Boltzmann equation*, J.Stat.Phys.23, 149-166(1980).

16. H.Cornille, *Oscillating and absolute Maxwellians: Exact solutions for (d>1)-dimensional Boltzmann equations*, J.Math.Phys.27, 1373-1386(1986).

17. H.Cornille,A.Gervois, *Power-like decreasing solutions of the nonlinear Boltzmann equation corresponding to Maxwellian interaction*, Phys.Lett.79A, 291-294(1980).

18. H.Cornille,A.Gervois, *Powerlike decreasing solutions of the Boltzmann equation for a Maxwell gas*, J.Stat.Phys.26, 181-217(1981).

19. P.Cvitanovic Editor, Universality in Chaos, A.Hilger Ltd, Bristol, 1984.

20. G.Dukek,T.F.Nonnenmacher, *Similarity solutions of the nonlinear Boltzmann equation generated by Lie group methods*, in Applications of Mathematics in Technology, V.Boffi and H.Neunzert Editors, 448-468, Teubner, Stuttgart, 1984.

21. G.Dukek,D.Rupp, *On the nonlinear spatially homogeneous Boltzmann equation with an external source: Exact generating functions for all moments*, Z.A.M.P. (J.Appl.Math.Phys.)37, 837-848(1986).

22. M.H.Ernst, *Nonlinear model Boltzmann equations and exact solutions*, Phys.Rep. 78, 1-171(1981).

23. M.J.Feigenbaum, *Universal behavior in nonlinear systems*, Los Alamos Science 1, 4-27(1980).

24. V.Franceschini,C.Tebaldi, *Sequences of infinite bifurcations and turbulence in a five-mode truncation of the Navier-Stokes equations*, J.Stat.Phys.21, 707-726(1979).

25. B.D.Ganapol,G.Spiga,D.H.Zanette, *An accurate evaluation of the distribution function in nonlinear extended kinetic theory*, to be published.

26. B.D.Ganapol,S.Oggioni,G.Spiga, *Evaluation of the distribution function for the N species Boltzmann equation in an infinite homogeneous medium*, in preparation.

27. B.S.Goh, Management and Analysis of Biological Populations, Elsevier-North-Holland, New York, 1980.

28. R.H.Goodwin, *A growth cycle*, in Socialism, Capitalism and Economic Growth, C. H.Feinstein Editor, 54-58, Cambridge Univ. Press, London 1967.

29. H.Grad, *Principles of the kinetic theory of gases*, in Handbuch der Physik, Vol.12, 205-294, Springer, Berlin, 1958.

30. J.Guckenheimer,P.Holmes, Nonlinear Oscillations, Dynamical Systems, and Bifurcations of Vector Fields, Springer, New York, 1983.

31. E.M.Hendriks,T.M.Nieuwenhuizen, *Solution to the nonlinear Boltzmann equation for Maxwell models for nonisotropic initial conditions*, J.Stat.Phys.29, 59-78(1982).

32. J.P.Holloway,J.J.Dorning, *The dynamics of coupled nonlinear model Boltzmann equations*, J.Stat.Phys.49, 607-660(1987).

33. E.Ikenberry,C.Truesdell, *On the pressures and the flux of energy in a gas according to Maxwell's kinetic theory*, J.Rat.Mech.Anal.5, 1-54(1956).

34. M.Krook,T.T.Wu, *Exact solutions of the Boltzmann equation*, Phys.Fluids20, 1589-1595(1977).

35. R.Lupini,S.Oggioni,G.Spiga, *Dynamics of a breeding process in a mixture of reacting gases*, in preparation.

36. R.Lupini,G.Spiga, *Chaotic dynamics of spatially homogeneous gas mixtures*, Phys. Fluids 31, 2048-2051(1988).

37. S.V.G.Menon, *Solution of the nonlinear Boltzmann equation for test particle diffusion with removal events*, Phys.Lett.113A, 79-81(1985).

38. R.G.Muncaster, *On generating exact solutions of the Maxwell-Boltzmann equation*, Arch.Rat.Mech.Anal.70, 79-90(1979).

39. A.A.Nikolskii, *The simplest exact solutions of the Boltzmann equation for the motion of a rarefied gas*, Sov.Phys.Dokl.8, 633-635(1964).

40. T.F.Nonnenmacher, *Application of the similarity method to the nonlinear Boltzmann equation*, Z.A.M.P.(J.Appl.Math.Phys.)35, 680-691(1984).

41. T.F.Nonnenmacher, *Exact similarity solutions for nonlinear particle transport in a host medium*, Trans.Th.Stat.Phys.15, 1007-1021(1986).

42. P.Richetti.A.Arneodo, *The periodic-chaotic sequences in chemical reactions: a scenario close to homoclinic conditions*, Phys.Lett.109A, 359-366(1985).

43. D.Rupp,G.Dukek, *Influence of various source terms on the relaxation process in a binary gas mixture*, in Rarefied Gas Dynamics 15, V.Boffi and C.Cercignani Editors, 75-84, Teubner, Stuttgart, 1986.

44. D.Rupp,G.Dukek,T.F.Nonnenmacher, *Particle transport in a host medium with an external source: exact solutions for a nonlinear homogeneous Boltzmann equation*, Z.A.M.P.(J.Appl.Math.Phys.)39, 605-618(1988).

45. D.Rupp,T.F.Nonnenmacher, *Solutions to the nonlinear Boltzmann equation for particle transport in a host medium*, Phys.Fluids 29, 2746-2747(1986).

46. G.Spiga, *Nonlinear problems in particle transport theory*, in Applications of Mathematics in Technology, V.Boffi and H.Neunzert Editors, 430-447, Teubner, Stuttgart, 1984.

47. G.Spiga, *On some problems in nonlinear transport*, in VII Congresso Nazionale A.I.M.E.T.A., Vol.1, 51-62, C.D.C., Udine, 1984.

48. G.Spiga, *A generalized BKW solution of the nonlinear Boltzmann equation with removal*, Phys.Fluids 27, 2599-2600(1984).

49. G.Spiga, *Dynamical systems in nonlinear transport theory*, Math.Rep.7/87, Univ. of Bari, Bari, 1987.

50. G.Spiga,R.L.Bowden,V.C.Boffi, *On the solution to a class of nonlinear integral equations arising in transport theory*, J.Math.Phys.25, 3444-3449(1984).

51. G.Spiga,T.F.Nonnenmacher,V.C.Boffi, *Moment equations for the diffusion of the particles of a mixture via the scattering kernel formulation of the nonlinear Boltzmann equation*, Physica 131A, 431-448(1985).

52. G.Spiga,P.Vestrucci,V.C.Boffi, *Dynamics of gas mixtures in an extended kinetic theory*, in Selected Problems of Modern Continuum Mechanics, W.Kosinski, T.Manacorda,A.Morro,T.Ruggeri Editors, 157-168, Pitagora, Bologna, 1987.

53. G.Tenti,W.H.Hui, *Some classes of exact solutions of the nonlinear Boltzmann equation*, J.Math.Phys.19, 774-779(1978).

54. C.Truesdell,R.G.Muncaster, Fundamentals of Maxwell's Kinetic Theory of a Simple Monoatomic Gas, Academic Press, New York, 1980.

55. G.Uhlenbeck,G.Ford, Lectures in Statistical Mechanics, American Physical Society, Providence, R.I., 1963.

56. L.Waldmann, *Transporterscheinungen in Gasen von mittlerem Druck*, in Handbuch der Physik, Vol.12, 295-514, Springer, Berlin, 1958.

57. U.Weinert,S.L.Lin,E.A.Mason, *Solutions to the nonlinear Boltzmann equation describing relaxation to equilibrium*, Phys.Rev.22A, 2262-2269(1980).

58. D.H.Zanette, *Two velocity gas diffusion with removal and regeneration processes*, Physica 148A, 288-297(1988).

59. D.H.Zanette,R.O.Barrachina, *Nonlinear particle diffusion in a time dependent host medium*, Phys.Fluids 31, 502-505(1988).

60. D.H.Zanette,C.R.Garibotti,R.O.Barrachina, *Power-law decreasing in solutions of the Boltzmann equation*, Phys.Lett.120A, 219-222(1987).

Lecture Notes in Mathematics

Edited by A. Dold, B. Eckmann and F. Takens

Editorial Policy

for the publication of proceedings of conferences and other multi–author volumes

---

Lecture Notes aim to report new developments – quickly, informally and at a high level. The following describes criteria and procedures for multi–author volumes. For convenience we refer throughout to "proceedings" irrespective of whether the papers were presented at a meeting.

The editors of a volume are strongly advised to inform contributors about these points at an early stage.

§ 1. One (or more) expert participant(s) should act as the scientific editor(s) of the volume. They select the papers which are suitable (cf. §§ 2 – 5) for inclusion in the proceedings, and have them individually refereed (as for a journal). It should not be assumed that the published proceedings must reflect conference events in their entirety. The series editors will normally not interfere with the editing of a particular proceedings volume – except in fairly obvious cases, or on technical matters, such as described in §§ 2 – 5. The names of the scientific editors appear on the cover and title page of the volume.

§ 2. The proceedings should be reasonably homogeneous i.e. concerned with a limited and well–defined area. Papers that are essentially unrelated to this central topic should be excluded. One or two longer survey articles on recent developments in the field are often very useful additions. A detailed introduction on the subject of the congress is desirable.

§ 3. The final set of manuscripts should have at least 100 pages and preferably not exceed a total of 400 pages. Keeping the size below this bound should be achieved by stricter selection of articles and NOT by imposing an upper limit on the length of the individual papers.

§ 4. The contributions should be of a high mathematical standard and of current interest. Research articles should present new material and not duplicate other papers already published or due to be published. They should contain sufficient background and motivation and they should present proofs, or at least outlines of such, in sufficient detail to enable an expert to complete them. Thus summaries and mere announcements of papers appearing elsewhere cannot be included, although more detailed versions of, for instance, a highly technical contribution may well be published elsewhere later.
Contributions in numerical mathematics may be acceptable without formal theorems resp. proofs provided they present new algorithms solving problems (previously unsolved or less well solved) or develop innovative qualitative methods, not yet amenable to a more formal treatment.
Surveys, if included, should cover a sufficiently broad topic, and should normally not just review the author's own recent research. In the case of surveys, exceptionally, proofs of results may not be necessary.

§ 5. "Mathematical Reviews" and "Zentralblatt für Mathematik" recommend that papers in proceedings volumes carry an explicit statement that they are in final form and that no similar paper has been or is being submitted elsewhere, if these papers are to be considered for a review. Normally, papers that satisfy the criteria of the Lecture Notes in Mathematics series also satisfy this requirement, but we strongly recommend that each such paper carries the statement explicitly.

§ 6. Proceedings should appear soon after the related meeting. The publisher should therefore receive the complete manuscript (preferably in duplicate) including the Introduction and Table of Contents within nine months of the date of the meeting at the latest.

§ 7. Proposals for proceedings volumes should be sent to one of the editors of the series or to Springer-Verlag Heidelberg. They should give sufficient information on the conference, and on the proposed proceedings. In particular, they should include a list of the expected contributions with their prospective length. Abstracts or early versions (drafts) of the contributions are helpful.

General Remarks

Lecture Notes are printed by photo-offset from the master-copy delivered in camera-ready form by the authors of monographs, resp. editors of proceedings volumes. For this purpose Springer-Verlag provides technical instructions for the preparation of manuscripts. Volume editors are requested to distribute these to all contributing authors of proceedings volumes. Some homogeneity in the presentation of the contributions in a multi-author volume is desirable.
Careful preparation of manuscripts will help keep production time short and ensure a satisfactory appearance of the finished book. The actual production of a Lecture Notes volume normally takes approximately 8 weeks.
For monograph manuscripts typed or typeset according to our instructions, Springer-Verlag can, if necessary, contribute towards the preparation costs at a fixed rate.

Authors of monographs receive 50 free copies of their book. Editors of proceedings volumes similarly receive 50 copies of the book and are responsible for redistributing these to authors etc. at their discretion. No reprints of individual contributions can be supplied. No royalty is paid on Lecture Notes volumes.
Volume authors and editors are entitled to purchase further copies of their book for their personal use at a discount of 33.3 %, other Springer mathematics books at a discount of 20 % directly from Springer-Verlag. Authors contributing to proceedings volumes may purchase the volume in which their article appears at a discount of 20 %.

Commitment to publish is made by letter of intent rather than by signing a formal contract. Springer-Verlag secures the copyright for each volume.